The Search for
Superstrings,
Symmetry, and
the Theory of
Everything

In Search of the Edge of Time

Hothouse Earth

Being Human

In Search of the Big Bang

In Search of Schrödinger's Cat

The Hole in the Sky

Stephen Hawking: A Life in Science
(WITH MICHAEL WHITE)

Albert Einstein: A Life in Science

The Matter Myth
(WITH PAUL DAVIES)

In the Beginning

Schrödinger's Kittens and
the Search for Reality

Companion to the Cosmos

John Gribbin

The Search for Superstrings, Symmetry, and the Theory of Everything

Little, Brown and Company

Boston   New York   London

FIRST NORTH AMERICAN EDITION

Portions of this book first appeared in *In Search of the Big Bang*

LIBRARY OF CONGRESS CATALOGING-IN-PUBLICATION DATA
Gribbin, John R.
    [In search of SUSY]
    The search for superstrings, symmetry, and the theory of
everything / John Gribbin. — 1st North American ed.
        p.   cm.
    Originally published: In search of SUSY. UK : Penguin, 1998.
    Includes bibliographical references and index.
    ISBN 0-316-32975-4
    1. Grand unified theories (Nuclear physics) 2. Particles
(Nuclear physics) 3. Supersymmetry.   I. Title.
QC794.6.G7G77   1999
530.14'2 — dc21   98-34711

10 9 8 7 6 5 4 3 2 1

MV-NY

Book design by Melodie Wertelet

Printed in the United States of America

Thanks to Benjamin Gribbin
for editorial assistance

Contents

Preface

When I revised my book *In Search of the Big Bang* to bring it up to date and tell the entire story of the life and prospective death of the Universe, something had to go to make way for the new material. That 'something' was mainly the detailed discussion of the world of sub-atomic particles, which was slightly tangential to the story of the Big Bang. No sooner had I done so, however, than various friends and colleagues that I discussed the project with bemoaned the loss, telling me that the kind of historical overview that the material had provided was all too rare in popular accounts of particle physics, or even in books aimed at students taking physics courses.

I took a second look, and felt that they might be right. So here, updated to the late-1990s, is the story of the particle world, from the discovery of the electron to the search for a supersymmetric theory explaining all of the forces and particles of nature in one mathematical package. It draws on materials from the original version of my Big Bang book, but does not overlap with the revised version of *In Search of the Big Bang*. The story isn't complete, because the mathematical physicists haven't yet found the ultimate theory of everything that they seek. But it will, I hope, shed some light on why they are looking where they are looking for the ultimate theory.

JOHN GRIBBIN

The Search for
Superstrings,
Symmetry, and
the Theory of
Everything

Introduction

The Material World

During the nineteenth century, chemists developed the idea, which dated back to the time of Democritus, in the fourth century BC, that everything in the material world is made up of tiny, indivisible particles called atoms. Atoms were thought of as being like tiny billiard balls, so small that it would take a hundred million of them, side by side, to stretch along a line 1 cm long. Atoms of a particular element each had the same mass, but the atoms of different elements, such as carbon, oxygen or iron, had different masses from one another, and the properties of the atoms, it was realized, determine the gross properties of larger quantities of the elements. When elements combine (for example, when carbon burns in air), it is because individual atoms of each element combine to make molecules (in this example, each atom of carbon combines with two atoms of oxygen to make carbon dioxide).

But just as the idea of atoms was becoming firmly established, in 1897 the English physicist J. J. Thomson, working at the Cavendish Laboratory in Cambridge, found a way to study bits that had been broken off atoms. The bits he broke off were much smaller and lighter than atoms, and carried negative electric charge; they were called electrons. They left behind 'atoms' with a residual positive charge, now known as ions. Thomson's experiments in the 1890s showed that although atoms of different elements are different from each other, they all contain electrons, and that the electrons broken off from any atom are the same as the electrons broken off from any other atom.

While physicists were still coming to terms with the idea that bits could be chipped off from the 'indivisible' atoms, the discovery of radioactivity was both giving them a new tool with which to probe the structure of atoms themselves and (although it was not realized at first) demonstrating that particles much larger than electrons could break off from atoms. At the beginning of the twentieth century, the New Zealander Ernest Rutherford, working at McGill University in Montreal with Frederick Soddy, showed that radioactivity involves the transformation of atoms of one element into atoms of another element. In the process, the atoms emit one or both of two types of radiation, named (by Rutherford) alpha and beta rays. Beta rays, it turned out, were simply fast-moving electrons. The alpha 'rays' also turned out to be fast-moving particles, but much more massive — particles each with a mass about four times that of an atom of hydrogen (the lightest element), and carrying two units of positive charge. They were, in fact, identical (apart from the speed with which they moved) to atoms of helium (the second lightest element) from which two electrons had been removed — helium ions. And their combination of relatively large mass (compared with an electron) and high speed gave Rutherford the tool he needed to probe the structure of atoms.

Soon Rutherford (by now working at the University of Manchester in England) and his colleagues were using alpha particles, produced by naturally radioactive atoms, as tiny bullets with which to shoot at the atoms in a crystal, or in a thin foil of metal. They found that most often alpha particles went right through a thin metal foil target, but that occasionally a particle would be bounced back almost the way it came. Rutherford came up with an explanation of this behaviour in 1911, and gave us the basic model of the atom that we learn about in school today.

Rutherford realized that most of the material of an atom must be concentrated in a tiny inner core, which he called the nucleus, surrounded by a cloud of electrons. Alpha particles, which come from radioactive atoms, are actually fragments of the atomic nucleus from which they are emitted (and are, in fact, nuclei of helium). When such a particle hits the electron cloud of an atom, it brushes its way through almost unaffected. But electrons carry negative charge, while atoms as a whole are electrically neutral. So the positive charge of an atom must be concentrated, like its mass, in the nucleus. Alpha particles too are positively charged. And when an alpha particle hits an atomic nucleus head on, the repulsion between like electric charges halts it in its tracks and then pushes it back from where it came.

Later experiments confirmed the broad accuracy of Rutherford's picture of the atom. Most of the mass and all of the positive charge is concentrated in a nucleus about one hundred thousandth of the size of the atom. The rest of the space is occupied by a tenuous cloud of very light electrons that carry negative charge. In round numbers, a nucleus is about 10^{-13} cm across,[1] while an atom is about 10^{-8} cm across. Very roughly, the proportion is like a grain of sand at the centre of Carnegie Hall. The empty hall is the 'atom'; the grain of sand is the 'nucleus'.

The particle that carries the positive charge in the nucleus is called the proton. It has a charge exactly the same as the charge on the electron, but with opposite sign. Each proton is about 2,000 times as massive as each electron. In the simplest version of Rutherford's model of the atom, there was nothing but electrons and protons, in equal numbers but with the protons confined to the nucleus, in spite of them all having the same charge, which ought to make them repel one another. (Like charges behave in the same way as like magnetic poles do in this respect.) As we shall see, there must therefore be another force, which only operates at very short ranges, that overcomes the electric force and glues the nucleus together. But over the twenty years following Rutherford's proposal of this model of the atom, a suspicion grew up among physicists that there ought to be another particle — a counterpart of the proton with much the same mass but electrically neutral.

---

[1] 10^{-13} means a decimal point followed by 12 zeros and a 1; 10^{13} means 1 followed by 13 zeros, and so on.

Among other things, the presence of such particles in the nucleus would provide something for the positively charged protons to hold on to without being electrically repulsed. And the presence of neutrons, as they were soon called, could explain why some atoms could have identical chemical properties to one another but slightly different mass.

Chemical properties depend on the electron cloud of an atom, the visible 'face' that it shows to other atoms. Atoms with identical chemistry must have identical numbers of electrons, and therefore identical numbers of protons. But they could still have different numbers of neutrons and therefore different masses. Such close atomic cousins are now called isotopes.

The great variety of elements in the world are, we now know, all built on this simple scheme. Hydrogen, with a nucleus consisting of one proton, and with one electron outside it, is the simplest. The most common form of carbon, an atom that is the very basis of living things, including ourselves, has six protons and six neutrons in the nucleus of each atom, with six electrons in a cloud surrounding the nucleus. But there are nuclei which contain many more particles (more nucleons) than this. Iron has 26 protons in its nucleus and, in the most common isotope, 30 neutrons, making 56 nucleons in all, while uranium is one of the most massive naturally occurring elements, with 92 protons and no less than 143 neutrons in each nucleus of uranium-235, the radioactive isotope which is used as a source of nuclear energy.

Energy can be obtained from the fission of very heavy nuclei because the most stable state an atomic nucleus could possibly be in, with the least energy, is iron-56. In terms of energy, iron-56 lies at the bottom of a valley, with lighter nuclei, including those of oxygen, carbon, helium and hydrogen, up one side and heavier nuclei, including cobalt, nickel, uranium and plutonium, up the other side. Just as it is easier to kick a ball lying on the valley's sloping side down into the bottom of the valley than to kick it higher up the slope, so if heavy nuclei can be persuaded to split, they can, under the right circumstances, form more stable nuclei 'lower down the slope', with energy being released. Equally, if light nuclei can be persuaded to fuse together, then they too form a more stable configuration with energy being released. The fission process is what powers an atomic bomb. The fusion process is what provides the energy from a hydrogen (or fusion) bomb, or of a star, like the Sun; in both cases hydrogen nuclei are converted into helium nuclei. But all that still lay in the future in the 1920s. Although there was circumstantial evidence for the existence of neutrons in that decade, it was only in 1932 that James Chadwick, a former student of Rutherford who was working at the Cavendish Laboratory (where Rutherford was by then the Director), carried out experiments which proved that neutrons really existed.

So the picture which most educated people have of atoms as being made up of three basic types of particles — protons, neutrons and electrons — really only dates back just over sixty-five years, less than a human lifetime. In that lifetime, things first got a lot more complicated for the particle physicists, and then began to get simple again. Those complications, and the search for a simplifying principle to bring order to the particle world, are what this book is all about. Many physicists now believe that they are on the verge of explaining the way all the particles and forces of nature work within one set of equations — a 'theory of everything' involving a phenomenon known as supersymmetry, or SUSY. The story of the search for SUSY begins with the realization, early in the twentieth century, that subatomic particles such as electrons do not obey the laws of physics which apply, as Isaac Newton discovered three centuries ago, to the world of objects such as billiard balls, apples, and the Moon. Instead, they obey the laws of the world of quantum physics, where particles blur into waves, nothing is certain, and probability rules.

Chapter One

Quantum Physics
for Beginners

Before 1900, physicists thought of the material world as being composed of little, hard objects — atoms and molecules which interacted with one another to produce the variety of materials, living and non-living, that we see around us. They also had a very good theory of how light propagated, in the form of an electromagnetic wave, in many ways analogous to the ripples on a pond or to the sound waves which carry information in the form of vibrations in the air. Gravity was a little more mysterious. But, by and large, the division of the world into particles and waves seemed clearcut, and physics seemed to be on the threshold of dotting all the *i*'s and crossing all the *t*'s. In short, the end of theoretical physics and the solution of all the great puzzles seemed to be in sight.

Scarcely had physicists started to acknowledge this cosy possibility, however, than the house of cards they had so painstakingly constructed came tumbling down. It turned out that the behaviour of light could sometimes only be explained in terms of particles (photons) while the wave explanation, or model, remained the only valid one in other circumstances. A little later, physicists realized that, as if waves that sometimes behave as particles were not enough to worry about, particles could sometimes behave like waves. And meanwhile Albert Einstein was overturning established wisdom about the nature of space, time and gravity with his theories of relativity. When the dust began to settle at the end of the 1920s, physicists had a new picture of the world which was very different from the old one. This is still the basis of the picture we have today. It tells us that there are no pure particles or waves, but only, at the fundamental level, things best described as a mixture of wave and particle, occasionally referred to as 'wavicles'. It tells us that it is impossible to predict with absolute certainty the outcome of any atomic experiment, or indeed of any event in the Universe, and that our world is governed by probabilities. And it tells us that it is impossible to know simultaneously both the exact position of an object and its exact momentum (where it is going).

How and why physicists came to these startling conclusions I have described at length in my book *In Search of Schrödinger's Cat*. Here I intend only to present an outline of the new world picture, without going into the historical and experimental details on which it is founded. But that foundation *is* secure; quantum physics is as solidly based, and as thoroughly established by experiments and observations, as Einstein's General Theory of Relativity. Together they provide the best description we have of the Universe and everything in it, and there is real hope that the two pillars of twentieth-century physics may yet be combined in one unified theory.

Photons

The best place to pick up the story of quantum physics and the search for unification is with the work of the great Scottish physicist James Clerk Maxwell, in the third quarter of the nineteenth century. Maxwell, who was born in Edinburgh in 1831, made many contributions to physics, but his greatest work was undoubtedly his theory of electromagnetism. Like many of his contemporaries and immediate predecessors, Maxwell was fascinated by the fact that an electric current flowing in a wire produces a magnetic field, which in its fundamentals is exactly the same as the magnetic field of a magnet itself. The field around a wire carrying a current will, for example, deflect a small compass magnet placed nearby. But also, a moving magnet, passing by a wire, will cause a current to flow in the wire. Moving electricity, a current, produces magnetism, and moving magnets produce electric currents. Electric forces and magnetic forces, which had once seemed to be quite separate phenomena, now seemed to be different facets of some greater whole, the electromagnetic field.

Maxwell tried to write down a set of equations that would link together all of the electric and magnetic phenomena that physicists had observed and measured. There were four equations: one to describe the magnetic field produced by an electric current, a second to describe the electric field produced by a changing magnetic field, the third giving the electric field produced by an electric charge itself, and the fourth giving a description of the magnetic field itself, including the strange fact that magnetic poles always come in pairs (north matched with south). But when Maxwell examined the equations, he found that they were flawed mathematically. In order to correct the maths, he had to introduce another term into the first equation, a term equivalent to a description of how a magnetic field could be produced by a changing electric field without any current flowing.

At that time, nobody had observed such a phenomenon. But once Maxwell had reconstructed the equations in the most elegant mathematical form, the reason for this extra term soon became clear. Physicists knew about condensers (now called capacitors), which are flat metal plates separated by a short gap across which an electric potential difference can be applied. One plate may be connected to the positive pole of a battery and the other plate to the negative pole. In this case, one plate builds up a charge of positive electricity, and the other a negative charge. The gap in between the plates is a region with a strong electric field, but no current flows across the gap and there is no magnetic field. Maxwell's new mathematical term described, among other things, what happens between the plates of such a capacitor just as the battery is connected to the plates. While the electric charge on the plates is building up, there is a rapidly changing electric field in the gap between the plates, and according to the equations this produces a magnetic field. Maxwell was soon able to confirm that the equations were correct, simply by placing a little compass magnet in the gap between two metal plates, and watching how it was deflected when the plates were connected to a battery. Like all the best scientific theories, the new theory of electromagnetism had successfully predicted how an experiment would turn out.

But now came the really dramatic discovery. Maxwell realized that if the changing electric field could produce a changing magnetic field, and the changing magnetic field could produce a changing electric field, the two components of the single, unified electromagnetic field could get along quite nicely together without any need for electric currents or magnets at all. The equations said that a self-reinforcing electromagnetic field, with the electricity producing the magnetism and the magnetism producing the electricity, could set off quite happily through space on its own, once it was given a push to start it going. The continually changing electromagnetic field predicted by the equations was in the form of a wave moving at a certain speed — 300,000 km/sec. This is exactly the speed of light. Maxwell's equations of electromagnetism had predicted the existence of electromagnetic waves

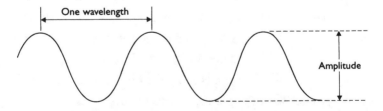

Figure 1.1   A wave is defined by its amplitude and its wavelength.

moving at the speed of light, and it didn't take Maxwell long to realize that light must indeed be an electromagnetic wave.

There was already a well-established body of evidence that light was a form of wave motion, so Maxwell's discovery fitted right in to the mainstream of nineteenth century science, and was welcomed with open arms. The best evidence for the wave nature of light comes from the way it can be made to 'interfere' with itself, like the interference between two sets of ripples on a pond, producing patterns of shade and light that cannot be explained in any other way.

Thomas Young, a British physicist and physician who was born in Somerset in 1773, produced the crucial experimental evidence in the early 1800s, when he shone a beam of pure light of one colour (monochromatic light) through a pair of narrow slits in a screen, to produce two sets of 'ripples' and make a classic interference pattern on a second screen. This work effectively pulled the rug from under the old idea, going back to Newton, that light came in the form of tiny particles, or corpuscles.

The combination of Maxwell's and Young's work provided what seemed to be a thorough understanding of light. Interference experiments made it possible to measure the wavelength of light, the distance from the crest of one wave to the next crest, which turns out to be about one ten-millionth (10^{-7}) of a metre; the different colours of the spectrum correspond to different wavelengths, with red light having a wavelength several times longer than blue, and Maxwell inferred that there must be electromagnetic radiation with a whole range of wavelengths extending far outside the visible spectrum, some much shorter than those of visible light and some much longer. Radio waves, with wavelengths of several metres, were produced by the German pioneer Heinrich Hertz before the end of the nineteenth century, and confirmed Maxwell's prediction.

All of the electromagnetic spectrum, from radio waves to visible light and on to X-rays, obeys Maxwell's equations. Those equations, describing how electromagnetic radiation propagates as waves, are the basis of the design of such familiar everyday objects as a TV set or a radio. They are also the

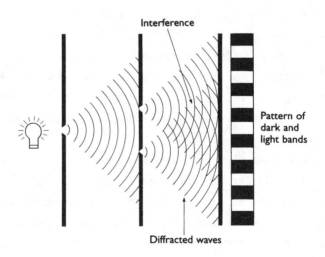

Figure 1.2   When two waves meet, they interfere with one another. This is shown clearly when a pure colour of light from a small source passes through two tiny holes in a screen. The two sets of waves spreading out from the two holes interfere to make a distinctive pattern of light and dark stripes on a second screen.

basis of the cosmological interpretation of the redshift, and, indeed, the notion of light as a wave is a firm and familiar concept. And yet since the early 1900s it has been clear that Newton was right all along. Light, and all forms of electromagnetic radiation, can be described in terms of particles, now called photons. In some circumstances, the behaviour of light can best be explained in terms of photons, as Einstein pointed out in 1905.

The first hints at the corpuscular nature of light had come in 1900, when Max Planck, a German physicist who had been born in Kiel in 1858, found that he was forced to introduce the idea of discrete packets of light into the equations which describe how light, or other electromagnetic radiation, is emitted by a hot body. This had been a major puzzle for physicists in the 1890s. They guessed that light was produced by the vibration of electrically charged particles inside an object, vibrations involving the atoms themselves, with the vibrations providing the push needed to set the waves described by Maxwell's equations off and running. And they knew, from observations and experiments, that the kind of radiation produced by an object depends on its temperature. We know this from everyday experience — a white-hot piece of metal (such as a poker) is hotter than a red-hot piece of metal, and a piece that is too cold to emit visible light at all may still be too hot to hold in your hand. Any such object radiates electromagnetic waves over a broad range of wavelengths, but the peak intensity of the radiation is always at a wavelength characteristic of the temperature of the object, and is shorter for hotter objects. The nature of the overall spectrum of radiation is always the same, and the position of the peak reveals very accurately the temperature of the radiation — for a 'perfect' radiator, it is the famous 'black body' spectrum. But until Planck came on the scene, nobody could manipulate the electromagnetic equations in such a way that they predicted the nature of the black body spectrum.

Planck found that the only way in which the observed nature of the black body spectrum could be explained was if light (by which I now mean any form of electromagnetic radiation) can only be emitted by the vibrating charges inside the atoms in little packets of energy.[1] By implication, that also meant that atoms could only absorb light in lumps of certain sizes. Planck expressed this in terms of the frequency of the radiation, denoted by the Greek letter nu, v. The frequency can be thought of as the number of wave crests passing a fixed point every second; for light with a wavelength of 10^{-7} metres and a velocity of 300,000 kilometres a second, the frequency is 3×10^{15} per second, or 3×10^{15} Hertz, in honour of the radio pioneer. Planck found that the observed black body spectrum could be explained if for every frequency of light there is a characteristic amount of energy equal to the frequency multiplied by a fundamental constant, which he called h. This energy, $E = hv$, is the smallest amount of energy of that

---

[1] The electron itself was only discovered, remember, in 1897, so Planck's explanation of black body radiation was necessarily a little vague on the exact nature of the charged particles within the atoms, and how they might be 'vibrating' to produce electromagnetic waves.

frequency that can be emitted or absorbed by any atom, and it can only emit or absorb quantities that are an exact multiple (1, 2, 3, 4 . . . n . . .) of this fundamental energy.

Planck did not suggest that the energy in the light only *existed* in little packets with energy $h\nu$; he thought that the restriction on the emission or absorption was something to do with the nature of the charged oscillators inside the hot objects. But he was able to calculate the value of h, which is the same for all radiation. It is now called Planck's constant, and it is tiny — 6.6×10^{-34} Joule seconds. Even for light with frequency 10^{15} Hz, the fundamental unit of energy is a mere 10^{-18} Joules, and it takes 6,000 Joules to keep a typical light bulb burning for one minute. The energy being radiated by a light bulb seems to be continuous because h is so small; in fact, the visible light is made up of billions upon billions of little packets of energy.

Planck's proposal met with a mixed reception at first. It seemed to explain the black body spectrum, but only by a kind of mathematical sleight of hand — a trick. Einstein, then an almost unknown physicist still working at his desk in the Swiss patent office, gave that mathematical trick a respectable physical reality when he showed that another great puzzle of the time could be explained if those little packets of energy had a real existence, and that light *only* existed in pieces with energy $h\nu$. And Einstein's attack on the puzzle of the nature of light provides a much clearer physical picture of why this must be so.

The photoelectric effect occurs when light shines on to a metal surface in a vacuum. The light literally knocks electrons out of the metal, and the electrons can be detected and the energy that they carry can be measured. The effect had been discovered in 1899 by the Hungarian Phillip Lenard. It was no great surprise to find that the energy in the light could make electrons jump out of the metal, but it *was* a great surprise to find just how the energy in the light and the energy in the electrons were related. Lenard used monochromatic light, so all the waves had the same frequency. A bright light obviously carries more energy than a dim light, so you might expect that if you shine a bright light on to a metal surface the electrons that are knocked out of it would each carry more energy. In fact, Lenard found that provided he used the same frequency of light it made no difference to the individual electrons how bright the light was. Each electron that jumped out of the metal always had the same amount of energy.

When Lenard moved the lamp closer to the metal, so that it shone more brightly on to the surface, there were indeed *more* electrons produced by the photoelectric effect, corresponding to the extra energy available from the brighter light. But each of those electrons carried the same amount of energy, and that was also the same amount of energy that each photoelectric electron carried when the light was dimmed, although there were fewer ejected electrons then. On the other hand, if he used light with a higher frequency (corresponding to shorter wavelength), the electrons produced had more energy, even if the light was dim. They still had the same energy as each other, but this was more than the energy of electrons produced by light with a lower frequency. The reason for all this seems simple with hindsight, but the suggestion Einstein made was revolutionary at the time. He simply suggested (and provided the mathematical calculations to back up the suggestion) that a beam of light of frequency v is made up of a stream of particles, what we now call photons, each of which has energy hv. An electron is ejected from the metal when one photon strikes one atom in the right way. So each ejected electron carries the amount of energy hv provided by one photon, less the amount of energy needed to tear the electron loose. The brighter the light, the more photons there are, so the more electrons are produced. But the energy of each photon stays the same. The only way to increase the energy of an individual photoelectron is to increase the energy of the photon that knocks it out of the metal, and the only way to do that is to make v bigger.

The suggestion was far from universally acclaimed by physicists at the time. Everybody knew, and Young's double-slit experiment and Maxwell's equations established this beyond reasonable doubt, that light was an electromagnetic wave. Only a brash newcomer with no real understanding of physics, it seemed, would dare to revive the preposterous Newtonian idea of light corpuscles. Indeed, one experimenter, the American Robert Millikan, was so incensed by the idea that he devoted ten years to experiments aimed at proving Einstein's hypothesis wrong. He succeeded only in proving it right, obtaining a very precise measurement of the value of h along the way, and helping to ensure that Planck received the Nobel Prize in Physics in 1918, that Einstein received the

1921 Prize for his explanation of the photoelectric effect, and that Millikan received the Prize in 1923. There is no doubt that all these awards were merited; the surprise is that Einstein never received a second Nobel Prize, for his General Theory of Relativity.[2]

By the time these honours came the way of the quantum pioneers, Planck's introduction of the quantum, *hv*, into atomic physics had helped other physicists, led by the Dane Niels Bohr, to develop the first satisfactory model of the atom — a model based on the idea, stemming from Rutherford's work, of a small, positively charged nucleus with even smaller, negatively charged electrons orbiting around it, more or less in the way the planets orbit around the Sun. The model said that these orbits were separated by intervals corresponding to a basic quantum of energy, and that an electron could jump from one orbit to another but could not exist in an in-between state. If it jumped from a higher energy orbit to a lower energy orbit, it would emit a photon with energy *hv*; to jump from a lower energy to a higher energy it had to absorb

---

[2] This is especially ironic since in recent years several theorists have pointed out that there *is* a way, after all, in which you can account for the photoelectric effect in terms of electromagnetic waves interacting with quantized atoms. Cunning though this trick is, however, it does not pull the rug from under quantum theory, since it uses modern quantum physics to explain how the atoms are behaving. And it is still true that, historically, it was the photoelectric effect that persuaded people that light came in quanta, as the award of Einstein's Nobel Prize shows. I'll stick with the historical account here.

a photon, *hv*. But because there was no such thing as half a photon, it could not jump to a state halfway between two of the allowed orbits.

The model was far from complete, but it gave physicists a handle on the way electrons behaved in atoms, and helped them to begin to explain the bright and dark lines of atomic spectra. Bright lines simply correspond to the emission of photons when an electron jumps down some rungs on the energy ladder; dark lines are the 'gaps' left in the spectrum when electrons absorb photons with a precisely defined frequency as they jump up the rungs of the ladder. But there were many questions still open in the early 1920s. For a start, the theory of electromagnetic radiation was not one theory, but two. Sometimes light, and X-rays, had to be described using Maxwell's wave equations; sometimes you had to use Einstein's photons; sometimes you had to use a mixture of the two ideas, as in Planck's calculation of the black body spectrum. And, the most pressing question of all, just what was it that decided which energy levels inside the atom could be occupied by electrons — what accounted for the number of rungs on the energy ladder and the spacing between them? The answers came not from any rationalization of the theory of light and a return to the calm logic of nineteenth century 'classical' physics, but from extending the revolution affecting waves to the world of particles. In particular, as Louis de Broglie suggested to a startled physics community in 1924, if light waves behave like particles, why can't electrons, which used to be thought of as particles, behave like waves?

Electrons

De Broglie, who was born in 1892, was the younger son of a French nobleman, and later inherited the family title from his brother to become the Duc de Broglie. His brother, Maurice, was a pioneer in the development of X-ray spectroscopy, and it was through Maurice that Louis became first aware of, and then fascinated by, the quantum revolution. The idea which he developed in his doctoral thesis, submitted to the Sorbonne in 1924, was brilliantly simple but backed up by thorough mathematical analysis. I shall only outline the simple, physical picture of de Broglie's insight into the nature of matter — if you want the maths, you will have to look elsewhere.

Einstein had developed the energy equation for material particles $E = mc^2$; Planck, with a little help from Einstein, had come up with a similar equation for photons, $E = h\nu$. Although a photon has no mass, it does carry momentum — if it didn't, it would not be able to knock electrons out of a metal surface. For an ordinary particle, the momentum is its mass (m) times its velocity (v). A light object moving very rapidly can carry as much energy, and give you as hard a knock, as a massive object travelling slowly. Think of the impact of a bullet compared with a softball. But the biggest knocks, of course, come from massive objects moving quickly. The velocity of a photon is c. And if we take out the factor (mc) from Einstein's equation and replace it by the letter p to represent momentum, we have a new equation, $E = pc$, which applies equally well to both ordinary matter and to photons.

So de Broglie put this equation and Planck's equation together. $E = pc = h\nu$. Rearrange things a little, and you have, for a photon, $p = h\nu/c$. But the velocity of a wave (c) divided by its frequency (ν) is just its wavelength. So de Broglie's version of the equation said that, for a photon, momentum is equal to Planck's constant divided by wavelength. This directly relates the particle nature of the photon (momentum) to its wave character (wavelength), using Planck's constant. But why, said de Broglie, stop there? Electrons have momentum, and we know the value of Planck's constant. Rearrange the equation again and we get a relation telling us that wavelength is equal to h divided by momentum. In other words, any particle, such as an electron, is also a wave, and its wavelength depends only on its momentum and on Planck's constant. For everyday particles, the mass involved, and therefore the momentum, is so big, compared with h, that the wave nature of matter can be ignored. For everyday objects, h divided by momentum is effectively zero. But for electrons, each with a mass of only 9×10^{-28} grams, the numbers are more nearly in balance, and the wave aspect becomes significant.

De Broglie suggested to his examiners that this strange equation had a physical reality, and that experiments might be carried out to measure the wavelength of electrons. His examiners were not sure whether to take this idea seriously, and regarded this aspect of his thesis presentation as more of a clever mathematical trick than anything of practical value. His thesis supervisor, Paul Langevin, sent a copy of the work to Einstein. It was Einstein who spotted at once the value of the work and its implications, and not only reassured Langevin that it was sound (so de Broglie was duly awarded his doctorate) but passed news of it on to other researchers. Within a few years, teams in both the United States and Britain had actually measured the wavelengths of beams of electrons, by scattering them off the atoms which form regular arrays in crystals. For an electron, this is the equivalent of Young's double-slit experiment, and it is just as conclusive.[3] Only waves can interact with each other to produce interference patterns, and under the right conditions electrons do just that. Meanwhile, the Austrian Erwin Schrödinger had developed a wave equation for the electron, equivalent to Maxwell's equations for light, which had proved to be one of the keys to developing a consistent model of the atom. De Broglie was duly awarded the 1929 Nobel Prize in Physics. By then, it was clear that all 'waves' have to be treated as 'particles', and that all 'particles' have to be treated as 'waves'. The confusion does not arise in everyday life, for objects big enough to see, or for waves on the ocean and ripples on the pond. But it is crucial to an understanding of atoms and subatomic phenomena. The physicists of the late 1920s were happy to have a consistent theory of the atom at last, even if the price they had to pay included some strange ideas about wave-particle duality. However, this strange aspect of atomic reality was almost the least

---

[3] One of the physicists involved was George Thomson, the son of J. J. In 1937, J. J. proudly saw his son receive the Nobel Prize for proving that electrons are waves, doubtless thinking back to that happy day in 1906 when he (J. J.) had received the Nobel Prize for proving that electrons are particles. Both awards were merited, and nothing better sums up the bizarre nature of quantum reality than this unusual father and son double.

of the strangenesses of the quantum world that was then un-
folding before their astonished gaze.

The Central Mystery

There is one experiment which contains the essence of quantum physics and lays bare its central mystery. This is the modern version of the two-slit experiment, which Young used to
prove that light is a form of wave. In practical terms, such an
experiment may be carried out using light, or electrons, or
other objects such as protons; it may not literally involve *two*
slits, but perhaps the equivalent of an array of slits, a so-called
diffraction grating, or the regularly spaced atoms in a crystal,
from which X-rays or electrons are bounced. But in order to
describe the central mystery in its purest form, I will talk in
terms of electrons, and of precisely two slits. And everything
that I will tell you about the behaviour of electrons in such circumstances has been checked and verified by experiments of
this kind, involving both electrons and photons and other
'particles'. None of this is just a bizarre hypothetical mathematical quirk of the equations; it is all tried and tested and
true.

The idealized form of the experiment is very easy to describe. It consists of a source of electrons (an electron 'gun',
like the one in the tube of your TV set), a screen with two
holes in it (small holes; they have to be small *compared with the
wavelength of an electron*, which is why the gaps between atoms
in a crystal turn out to be just right), and a detector. The de-

tector might be a screen, like a TV screen, which flashes when an electron hits it. What matters is that we have some means of recording when and where an electron hits the detector, and of adding up the number of electrons arriving at each place on the detector screen. When waves go through such a system, each of the two holes in the first screen becomes a source of waves, spreading out in a semicircle and marching in step with the waves from the other hole. Where the waves add up, they produce large vibrations; where they cancel out, there is no detectable vibration at all. This is why such an experiment with light waves produces bands of dark and light stripes on the detector screen.

If there is only one hole in the first screen, of course, then there is simply a bright spot (or band) of light on the detector, brightest at the centre and fading out smoothly on either side.

The same sort of thing happens when we fire a beam of electrons through a screen with *one* hole (one slit). The detector screen records most flashes in line with the hole, and a few on either side of the region where the electron beam is most intense. With our two-slit experiment, we can test this by blocking up first one hole and then the other. In each case, one hole on its own produces a bright spot on the screen which fades away smoothly on either side. But when *both* holes are open, then there is a clear diffraction pattern on the screen. The flashes which mark the arrival of individual electrons form bright stripes separated by dark regions. This is explained by the wave-like nature of electrons. The electron waves going through the two holes are interfering with one another, cancelling out in some places and reinforcing in others, just like light waves.

So far, so good. It is more than a little strange that electrons can behave like waves when they are going through the experimental apparatus, then suddenly coalesce into hard little lumps to strike flashes from the detector screen, but by combining the ideas of particle and wave we can at least begin to convince ourselves that we have some idea of what is going on. After all, a water wave is actually made up of myriads of little particles (water molecules) moving about. If we are firing hundreds of thousands of electrons in a beam through the two holes, perhaps it isn't so surprising that they can be guided in some way like waves while retaining their identity as little particles. If we fire just *one* electron at a time through the experiment, then logically we would expect that it would go through one hole or the other. The diffraction pattern we observe is simply, according to everyday logic, the result of observing many electrons at the same time.

So, what happens when you do fire one electron at a time through the experiment? Clearly, when you get one flash on the screen on the other side that doesn't tell you much about how the electron has behaved. But you can repeat the single shot experiment time after time, observing all the flashes and noting down their positions on the screen. When you do this, you find that the flashes slowly build up into the old, familiar diffraction pattern. Each individual electron, passing through the apparatus, has somehow behaved like a wave, interfering with itself and directing its own path to the appropriate bright region of the diffraction pattern. The only alternative explanation would be that all of the electrons going through the apparatus at different times have interfered with each other, or the 'memory' of each other, to produce the diffraction pattern.

It looks as if each electron goes through *both* slits. This is crazy. But we can devise an additional set of detectors which takes note of which slit each electron goes through, and repeat our experiment to see if that is indeed what happens. When we do this, we do not find that our detectors at the two holes each report the passage of an electron (or half an electron). Instead, sometimes the electron goes through one slit, and sometimes through the other. So what happens now when we send thousands of electrons through the apparatus, one after the other? Once again, a pattern builds up on the detector screen. But it is *not* a diffraction pattern! It is simply a combination of the two bright patches we get when one or the other of the holes is open, with no evidence of interference.

This is *very* strange. Whenever we try to detect an electron, it responds like a particle. But when we are not looking at it, it behaves like a wave. When we look to see which hole it goes through, it goes through only one hole and ignores the existence of the other one. But when we don't monitor its passage, it is somehow 'aware' of both holes at once, and acts as if it had passed through them both.

Quantum physicists have some nice phrases to describe all this. They say that there is a wave of some sort associated with an electron. This is called the 'wave function', and it is spread out, in principle, to fill the Universe. Schrödinger's equation describes these wave functions and how they interact with one another. The wave function is strongest in one region, which corresponds to the position of the electron in everyday language. It gets weaker further away from this region, but still exists even far away from the 'position' of the electron. The equations are very good at predicting how particles like electrons behave under different circumstances, including how they will interfere with one another when they, or the wave functions, pass through two slits. When we look at an electron, or measure it with a particle detector, the wave function is said to 'collapse'. At that instant, the position of the electron is known to within the accuracy allowed by the fundamental laws. But as soon as we stop looking, the wave function spreads out again, and interferes with the wave functions of other electrons — and, under the right circumstances, with itself.[4]

All of this is precisely quantifiable mathematically, and makes it possible to calculate how electrons fit in to atoms, how atoms combine to make molecules, and much more besides. The jargon term 'collapse of the wave function' (which has a precise mathematical significance in quantum theory) is equivalent to saying that we can only know where things are when we are actually looking at them. Blink, and they are

---

[4] How do we know what the electron wave does when we are not looking at it? From repeated observations in many experiments of where the electrons are, and which way they are moving when we do see them, physicists infer that the correct mathematical description of the wave function is in line with this very simple description. But if you want the details, then you will have to check them out in *In Search of Schrödinger's Cat*.

gone. And the behaviour of the particles depends on whether or not we are looking. If we watch the two holes to see electrons passing by, the electrons behave differently from the way they behave when we are not looking. The observer is, in quantum physics, an integral part of the experiment, and what he or she chooses to watch plays a crucial role in deciding what happens.

The implications of all this are very deep indeed. For one thing, we can no longer say that an electron, in principle identifiable as a unique object, starts at one side of our experiment and follows a unique path — a trajectory — through to the other side. The very concept of a continuous 'trajectory' is a hangover from classical Newtonian ideas, and has to be abandoned. Instead, quantum physicists talk in terms of 'events', which may happen in a certain order in time but which tell us nothing about what happens to the particles involved in events when they are not being observed. All we can say is that we observe an electron at the start (event 1), and that we observe an electron at the finish (event 2). We can say nothing at all about what it does in between, and indeed we cannot say that it is the same electron that is recorded at each event. Fire two electrons off together, and although two electrons arrive on the detector screen a little later, there is no way of telling which one is which.

Electrons are indistinguishable from one another, in a far deeper sense than any mass-produced objects of the everyday world, such as paperclips, are indistinguishable from one another. The electrons in an atom are not physically distinct entities, each following its own well-defined orbit. Instead, all we can say is that a particular kind of atom behaves as if it had associated with it a combination of eight, or ten, or whatever the number might be, electron wave functions. If we carry out an experiment designed to prod the atom (perhaps by bombarding it with photons, as in the photoelectric experiment), one or more of the electron wave functions may be modified in such a way that there is a high probability that we will detect an electron outside the atom, as if a little particle had been ejected. But the only realities are what we observe; everything else is conjecture, hypothetical models which we construct in our minds, and with our equations, to enable us to develop a picture of what is going on.

Which is more real, the particle or the wave? It depends on what question you ask of it. And no matter how skilful a physicist the questioner may be, there is never any absolute certainty about the answer that will come back.

Chance and Uncertainty

A particle is something that is well defined. It exists at a point in space, it occupies a small volume and has some kind of tangible reality, in terms of our everyday experience of the world. But a wave is almost the opposite. A pure wave stretches on forever, so there is no sense in which it can be said to exist at a point. It may have a very well defined direction — it carries momentum. But there is no way, even in imagination, that you can put your finger on it and hold it still while you look at it. So how can the two aspects of the subatomic world be reconciled?

In order to express itself in particle terms — as a photon, or as an electron — a wave must be confined in some way. Mathematicians know all about this. The way to confine a wave is to reduce its purity. Instead of a single wave with one unique, well-defined frequency, think of a bundle of waves, with a range of frequencies, all moving together. In some places the peaks of one wave will combine with the peaks of other waves to produce a strong wave; in other places the peaks of one wave will coincide with the troughs of other waves, and they will cancel each other out. Using a technique called Fourier analysis, the mathematicians can describe combinations of waves which cancel out almost completely everywhere except within some small, well-defined region of space. Such combinations are called wave packets. In principle, as long as you include enough different waves in the packet you can make it as small as you like. Since mathematicians use the Greek letter delta (Δ) to denote small quantities, we can say that the length of the wave packet is Δx. By losing the purity of a single wave with a unique frequency, we can localize the wave packet until it has the dimensions of an electron.

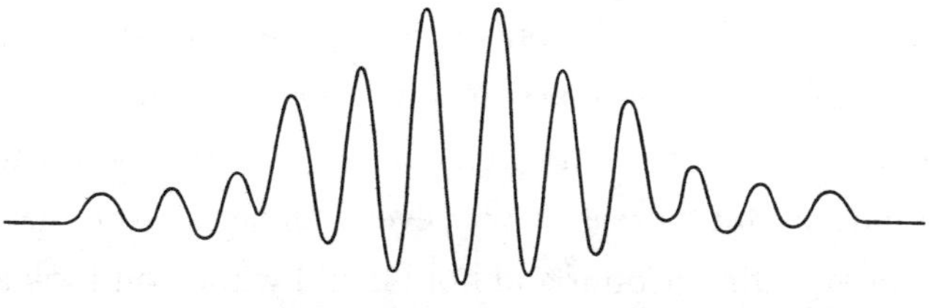

Figure 1.3   A wave packet is a group of waves that covers only a definite region of space.

But we have paid a price. The momentum of a wave, as de Broglie showed, is *h* divided by the wavelength. A pure wave has a unique wavelength, and therefore it has a unique momentum. But by introducing a mixture of wavelengths, or frequencies, we have introduced a mixture of momenta. The more waves we have (the smaller the wave packet), the less precise is the meaning of the term 'momentum' for this wave packet. All we can say is that there is a range of momenta, covering an amount Δp. Δx is the amount of uncertainty in the position of the wave packet; we know it is somewhere within a volume Δx across. Δp is the uncertainty in the momentum of the wave packet. We know roughly where the packet is going, but only to a precision of Δp. It is very simple, mathematically, to show that we can never reduce Δx or Δp to zero, and that, in fact, the *product* of the two uncertainties, $\Delta x \Delta p$, is always bigger than Planck's constant, *h*, divided by twice the value of pi (2π). This slightly different constant is written $\hbar$ and is either referred to as 'h cross' or, more sloppily, as 'Planck's constant', even though it is really $h/2\pi$. The relationship between position and momentum is Heisenberg's uncertainty relation, named after the Nobel Prize-winning German physicist Werner Heisenberg, who was one of the pioneers who developed quantum mechanics in its first full form in the 1920s, and is written:

$$\Delta x \Delta p > \hbar$$

Now, it is impossible to stress too strongly the fact that this relationship, this equation, is not just some weird mathematical trick. The empirical evidence for wave-particle duality means that it is impossible, in principle, to measure both the position of a particle and its momentum with absolute precision. Indeed, if you could measure *exactly* where an electron was, so that Δx is zero, then Δp would become infinite, and you would have no idea at all where in the Universe the electron would pop up next. And the uncertainty is not restricted solely to our *knowledge* of the electron. It is there all the time, built in to the very nature of electrons, and other particles and waves. The particle itself does not 'know', with absolute precision, both where it is now and where it is going next. The concept of uncertainty is intimately linked with the concept of chance in quantum physics. We cannot be certain where a particle is, and we cannot be certain where it is going, so we must not be too surprised if it turns up where we don't expect it.

Position and momentum are not the only properties of a particle that are related in this way. There are other pairs of what are called 'conjugate variables' that are similarly linked by the wave equations, and the most important of these are energy (E) and time (t). It turns out, from a rigorous mathematical analysis, that there is also an inherent uncertainty in the amount of energy involved in a subatomic process. If energy is transferred from one particle to another, and if the transfer takes a certain amount of time (which it must do, since nothing travels faster than light), then the uncertainty in the energy (ΔE) multiplied by the uncertainty in the time (Δt) is also bigger than h cross:

$$\Delta E \Delta t > \hbar$$

For a short enough time, indeed, both a particle and its immediate surroundings — or indeed the whole Universe — may be uncertain about how much energy the particle has. The strange 'tunnel effect', by which alpha particles get out of atomic nuclei, graphically illustrates the power of uncertainty in the subatomic world. George Gamow, in the 1920s, explained alpha emission 'properly', using the full equations of quantum physics. But we can easily see what is going on in general terms.

The alpha particle sits inside the nucleus, and we can imagine it as being just inside the rim of a volcano. If the particle were just *outside* the rim, it would 'roll away', and be ejected by the force of electric repulsion. The 'distance' from the inside of the nucleus to the outside is Δx. An alpha particle associated with the nucleus has a very well defined momentum, as does the nucleus itself. But that means its position must be uncertain. Even though an individual alpha particle does not have enough energy to climb the inner rim of the volcano and escape, it is not *inside* the nucleus, in the everyday meaning of 'inside'. Uncertainty implies that there is a finite, and precisely calculable, chance that the particle is actually outside the nucleus. Bingo! Some particles do find themselves outside the nucleus, take note of the fact, and rush away, just as if they have 'tunnelled' through the intervening barrier. It is exactly as if you took some dice and rattled them in a cup until suddenly one of them appeared outside the cup, rolling across the table. And if Planck's constant were big enough, that is how dice would indeed behave in the everyday world.

Or think of it in terms of energy. The particle needs more

energy to climb over the 'rim' of the potential barrier. The extra energy it needs is ΔE. For a brief enough instant of time, Δt, it might, for all the laws of physics know or care, have that extra energy. And if it does, once again it is off and running. It doesn't matter that it has to 'give back' the energy it borrowed from uncertainty, after a time Δt, because by then it has escaped down the hill on the other side of the barrier.

As if all this were not mind-bending enough, the quantum theory has more tricks in store. So far, we have looked mainly at the implications of a mixture of wave characteristics for particles. What happens to our understanding of waves, especially light, when we have to treat them as particles? The kind of problem the physicists encounter can be highlighted by one example. There are materials, called polarizers, which only al-

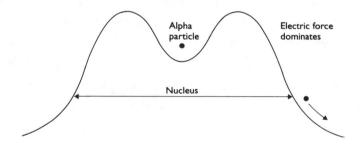

Figure 1.4 Quantum mechanical uncertainty explains how an alpha particle can escape from an atomic nucleus. It can 'borrow' a small amount of energy from uncertainty to climb over the hill before falling down the other side and giving up the energy as it makes its escape; or, from another point of view, the uncertainty in its position allows it to appear outside the barrier, as if it had tunnelled through the hill, and roll away to lower energies.

low light to pass through them if the waves are vibrating in a certain direction. Some sunglasses work like this. Since daylight contains waves vibrating in all directions, glasses that let only some of the waves through reduce glare. A 'polarized' light wave is one which has been through such a filter, and so all the waves that remain are vibrating in the same plane, up and down, say, if that is the way the polarizer was set up. If such a wave meets another polarizer aligned exactly across its own plane of polarization, none of the light gets through. But if the second polarizer is aligned at an angle less than 90° to the plane of polarization of the light, some of it gets through. The fraction that gets through depends on the angle; when it is 45°, exactly half the energy in the light gets through the second polarizer, as a wave polarized at 45° to the polarization of the first wave.

Maxwell's equations can explain all this. A polarizer at 45° takes out half the energy in the wave; a polarizer at right angles takes out all the energy. But what happens to individual photons coming up to the polarizer? You cannot chop a photon in half. It is the fundamental, basic unit of energy. So, in a beam of light passing through a polarizer at 45°, half the photons get through and half do not. But how are they selected? By chance, at random in accordance with the statistical rules of probability. When an individual photon arrives at the polarizer there is, in this example, a precise 50:50 chance that it will get through or that it will be stopped. The numbers change with the angle of the polarizer, but the principle does not. Individual photons are selected by the polarizer on a basis of pure chance. And this example is just a simple demonstration of what is happening throughout the quantum world. Every time subatomic particles are involved in interactions the outcome depends on chance. The odds may be very heavily stacked in favour of one particular outcome, or they may be no better than tossing a coin on a 50:50 basis. But they are clearly and precisely laid down by the laws of quantum physics, and there is no such thing as certainty in the quantum world.

This is the point about quantum theory which made Albert Einstein reject the whole thing, with his famous remark about God, 'that He would choose to play dice with the world . . . is something that I cannot believe for a single moment' (often paraphrased as 'I cannot believe that God does play dice'). But all the evidence is that God *does* play dice. Every experiment confirms the accuracy of the quantum interpretation. When we carry out an experiment, which might involve measuring the position of an electron, for example, we cannot know for certain how things are going to develop later. In this simple case we can say, perhaps, that there is a certain probability that next time we look the electron will have moved from point A to point B, a different probability that it will be at point C, and so on. The probabilities can all be calculated, in principle, and it might be far more than 99 per cent probable that the electron will go from A to B. But it is never *certain*. One day, when you do the old familiar experiment in which the electron goes from A to B, it might, just by chance, turn up at point C, or D, or Z, instead.

In the everyday world, we are saved from the more bizarre possibilities by the sheer numbers of particles involved. Billions and trillions of electrons move through the circuitry to make my wordprocessing computer work. A few mavericks somewhere in the system may indeed be blithely hopping from A to D instead of from A to B, but the vast majority do what they should, as far as I am concerned. Unless you have a taste for philosophy, you don't have to worry too much about the uncertain and chancy aspects of quantum theory in your own life. Even if, like me, you *do* have a taste for metaphysics, you can still use a computer without any *real* fear that *all* of the electrons will suddenly stop obeying orders. But the stranger aspects of quantum physics become very important when we deal with the subatomic world. We need just one more fundamental concept, and a few bizarre tidbits from the quantum cookbook, before we are, at last, ready to tackle those puzzles.

Path Integrals and a Plurality of Worlds

The fundamental difference between quantum mechanics and classical (Newtonian) mechanics is very clearly brought out when we look closely at the way a particle, such as an electron, actually does get from one point (A) to another (B). In the classical view, a particle at point A has a definite speed in a definite direction. As it is acted upon by external forces, it moves along a precisely determinable path, which, for the sake of argument, passes through, or ends at, point B. The quantum-mechanical view is different. We *cannot* know, not even in principle, both the position and momentum of a particle simultaneously. There is an inherent uncertainty about where the particle is going, and if a particle starts out at point A and is later detected at point B we cannot know exactly how it got from A to B, unless it is watched all the way along its path.

Richard Feynman, a Nobel Prize-winning physicist from the California Institute of Technology, applied this quantum-mechanical view to the history of particles as presented in the kind of space-time diagrams used by relativists. These are diagrams like graphs, with one axis representing time and the other space. Curves on the diagram (world lines) represent particle trajectories or histories, some of which are ruled out because they would involve travel faster than light, but many of which indicate valid ways for a particle to get from A to B. Going back to the experiment with two holes, for example, you might imagine literally drawing out a map of all the pos-

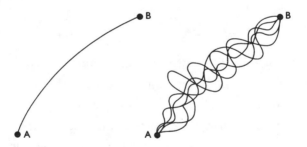

Figure 1.5 Classical (that is Newtonian) physics says that a particle follows a definite trajectory from A to B. Richard Feynman's quantum mechanical 'sum over histories' approach says that we must calculate the contribution of every possible path and add them together. Also known as the 'path integral' approach, this explains, among other things, how a single electron (or a single photon) can pass through both holes in a double-slit experiment (see *Figure 1.2*) and interfere with itself.

sible ways in which an electron could get from the gun on one side of the screen to a particular spot on the detector screen, passing through one or other of the holes on the way. Some of these trajectories, or paths, are very straightforward and direct; others meander about. Feynman's maps include time as well as space, and on them some of the trajectories represent fast passages of an electron through the experiment, and others represent slow passages. But each path, direct or circuitous, fast or slow, has associated with it a definite probability (strictly speaking a 'probability amplitude'), which can be calculated. The amplitudes are measured in terms of a quantity called action, which is energy × time, and which happens to be the unit Planck's constant, h, is measured in.

The probabilities of the world lines are not all 'in step' with one another, and like the amplitudes of ripples on a pond they can interfere with one another to reinforce the strength of one path while cancelling out the amplitudes of others. It is not unlike the way in which the waves in a wave packet cancel each other out everywhere except in a small region, Δx. Feynman's work[5] shows that when all of the amplitudes corresponding to possible particle trajectories are added together, the result of the interference is to wipe out all of the possible contributions except the ones which are very close to the trajectory that corresponds to the path from A to B in accordance with classical mechanics. And when the technique is

---

[5] Notably described in all its technically detailed glory by Feynman and A. R. Hibbs in their book *Quantum Mechanics and Path Integrals*, McGraw-Hill, 1965.

applied to experiments like the two-slit experiment, the results it comes up with are exactly the same as the results you get using Schrödinger's wave equation.

Actually, the technique has only been used in fully worked out detail for some very simple, special cases. Imagine the complexity of calculating the probabilities of every path from the electron gun to *each* point on the detector screen. The number of paths involved is so enormous in most cases that it is quite impractical to apply Feynman's technique in its pure form. But the concept underlying this approach, and the fact that it can be shown mathematically to make the same predictions as Schrödinger's version of quantum physics, are fundamentally important.[6] Feynman tells us that in the two-slit experiment we not only have to think of the electron as going through both holes at once, but as taking *every possible path* through both holes at once. The conventional quantum view has it that there is *no* trajectory; from Feynman's point of view, we have to take account of *every* trajectory.

This approach to describing the trajectories of particles is called the 'path integral' technique, because it involves adding

---

[6] There are, however, ways to generalize the technique, and to calculate its broad implications without calculating every path in detail. Feynman has proved, for example, that the most probable path, which corresponds to the classical trajectory, follows a line of least action, and he has established mathematically that only paths close to the line of least action need be included in the calculations, since the probability amplitudes from the other paths must cancel each other out. The problem with the two-slit experiment is that there are paths through each of the holes which each have the same 'least action', and all such paths are equally important in calculating the fate of the electron.

up different possible paths, or sometimes, more grandiosely, the 'sum over histories' approach. The alternative name echoes the idea behind an interpretation of the quantum rules which is far from being the majority view today, but which I am fond of and which I discussed in *In Search of Schrödinger's Cat*. The model fits the 'many worlds' description of reality developed originally by Hugh Everett of Princeton University in 1957, and recently enthusiastically espoused by David Deutsch in Oxford.[7]

What Everett found was that the equations could be interpreted, with complete validity, as implying that every time the Universe is faced with a 'choice' at the quantum level it splits into two and both options are chosen. In an experiment where an electron goes from point A to point B via an intervening screen which has two holes in it, quantum theory says that unless we watch all the time we cannot possibly tell which hole it went through, indeed that it is meaningless to say it went either way. Its 'real' trajectory is given by a sum of the two possible paths. But classical theory says there is a definite path and it must have gone through just one of the holes, even if we weren't looking. When we look to see which hole the electron goes through, of course, that particular uncertainty vanishes and we have a different experiment in which we know which path the particle took. But, says Everett (or rather, say the equations), for every observer who looks and sees the electron go through one hole, there is another ob-

---

[7] See his book *The Fabric of Reality*, published in 1997 by Allen Lane.

server — in another world — who looks and sees it go the other way. Both are equally real. Or, in terms of the photon passing through a piece of polarizer, every time a photon is faced with the 50:50 choice described above, the Universe divides into two. In one universe, the photon passes the filter; in the other universe, it does not. The strangest thing about this version of quantum reality is that it makes exactly the same predictions as the probabilistic interpretation for the observable outcomes of all experiments that could be carried out. This is both a strength of the model (after all, to be a good model it must agree with all those experiments done so far) and a weakness, since most theorists say with relief that since it makes no new, testable predictions to distinguish itself from the conventional interpretation, in that case we don't need it at all, and can stick with the probabilities. Indeed, probability provides the means to retain an image of the electron as a point-like particle, if you really want to retain that image.

Out of the Frying Pan?

English physicist Paul Davies, who is based at the University of Adelaide in Australia, urges this point of view in an under-graduate textbook he wrote on quantum physics.[8] 'Resist at all costs', he says in that book, 'the temptation to think of an electron as pulled asunder and smeared out in space in little ripples. The electron itself is not a wave. Rather, the way it moves about is controlled by wave-like principles. Physicists

---

[8] *Quantum Mechanics*, Routledge & Kegan Paul, London, 1984.

still regard the electron as a point-like entity, but the precise location of that point may not be well defined.' And he goes on to describe the probability waves that determine where an electron is likely to turn up next by making an analogy with a crime wave. 'Crime waves are not waves of undulating stuff but *probability* waves . . . crime waves, like fashions or unemployment, may move about — they have dynamics — but an individual crime still occurs, of course, at a place. It is the abstract probability which moves.'

For many purposes, and especially for teaching undergraduate physics, physicists do indeed treat electrons as 'real' particles, and the waves associated with them as 'probability waves', which can interfere with one another, be diffracted through small holes, and do all the other tricks waves can do. 'It is the probability which has the wave-like behaviour,' Davies tells his students, 'while the particles themselves remain as little lumps, albeit elusively secreted in the wave which guides their progress . . . which facet of this wave-particle duality is manifested depends on the sort of question that is asked.' This is bad teaching. You might try, if only he were still alive, asking Maxwell how he felt about the suggestion that light waves are only probability waves guiding the motion of little hard lumps called photons. His response might be interesting. No matter how hard you try to hold on to the image of an electron, or a photon, or whatever, as a particle, the concept persists in slipping away.

Take spin, for example. Electrons, and other subatomic particles, have a property which physicists call spin. It is fundamentally important in deciding how electrons are arranged in atoms, among other things, and it is measured in the same kind of units as the spin of a top, or of the Earth rotating in space. But there the analogy stops. The spin of an electron can only point in one of two directions — 'up' or 'down' — never 'sideways' or anywhere in between. Spin itself is quantized, like energy. The spin of a fundamental particle is measured in units of h cross. In those units, the spin of an electron is ½ — either +½ or –½, but never anything else. All the particles that we like to think of as 'real' particles, like protons, neutrons and electrons, have spin which is half-integral — ½, ³/₂, ⁵/₂ and so on.

All such particles obey a set of statistical rules known as Fermi-Dirac statistics, and are known as fermions. The photon, which has spin one, and all particles with integer spin (1, 2, 3 and so on), obey a different set of rules called Bose–Einstein statistics, and are known as bosons; so there *is* a fundamental difference between photons and electrons.

The most important distinction is that particles with half-integer spin like electrons (fermions) are exclusive. In terms of the 'ladder' of energy levels inside an atom, this means that only two electrons can sit on each rung of the ladder — one with spin up, the other with spin down.[9] A third electron is excluded from that rung because it would occupy an identical

---

[9] I am, in fact, oversimplifying. For reasons which I won't go into here, there are some sets of energy levels inside an atom where four 'rungs' of the ladder, or one rung on each of four separate ladders, in effect lie side by side, so eight electrons can have very similar states to one another. The relevant point, however, is still that strictly speaking no two of these electrons are in *identical* states.

state to one or the other of the two electrons already there. Photons and other integer-spin particles (bosons) are less snobbish. You can pack them in any old how, any old where. Furthermore, although fermions are conserved (overall, the number of fermions in the Universe stays the same), bosons are more ephemeral. We can make photons just by turning on a light; they vanish again when they are absorbed by atoms and give up their energy.

All of this is rather hard to reconcile with the existence of 'real' little lumps being guided by probability waves. It is even harder when the physicists tell you that the spin of a fundamental particle has some other peculiar properties. If you think of an electron as rotating, for example, it has to rotate through 360° not once but *twice* in order to get back to where it started. And although I just told you that fermions are conserved in the Universe at large, that constraint still allows you to make particles and antiparticles in equal numbers, if you have the energy available to do so. An electron-positron pair, for example, counts as zero in the total number of fermions on the Universe's list. The particle and antiparticle cancel out. If you've got the energy, you can make electron-positron pairs, just as they were made in the Big Bang. Where can you get the energy from today? If you are being mundane, you might imagine getting it by smashing particles together in giant accelerators, like the one at CERN. But you can be more imaginative. The limits of uncertainty principle allow you, if you do it quickly enough, to 'borrow' enough energy from the uncertainty relation to make particles, as long as they disappear again when their allowed time is up.

Take electrons. If the mass of an electron is m, then the energy needed to make an electron-positron pair is $2mc^2$. This is about one million electron Volts (1 MeV) in the units usually used by particle physicists. The laws of quantum physics allow such a pair of particles to pop into existence out of the vacuum for a very, very tiny split second of time (Planck's constant divided by 1 MeV), and then to annihilate one another and disappear again. Such particle pairs are called 'virtual' particles. Each pair can only exist for a very short time, but the vacuum is seething with such pairs, constantly being produced, disappearing, and being replaced by new pairs. At least, that is what quantum physics says the vacuum is like. And the existence of virtual particles has a direct effect on the equations of particle physics. Without virtual particles, the equations do not predict correctly the interactions between charged particles. With effects due to virtual particles included, they do.

So just how 'real' are any of the particles in the Universe? When Paul Davies is talking to his fellow researchers, instead of to his undergraduates, he takes a rather different line. In his contribution to a book published to mark the sixtieth birthday of physicist Bruce DeWitt (one of the champions, incidentally, of the many worlds interpretation of quantum physics), Davies presented an essay provocatively titled 'Particles Do Not Exist'.[10] The nub of the argument he put forward, one which agrees with the view of many theorists, starts from the fact that we cannot see, touch or feel fundamental particles

---

[10] *Quantum Theory of Gravity*, Adam Hilger, Bristol, 1984.

such as electrons. All we can do is carry out experiments, make observations, and draw conclusions about what is going on based on those observations and our experiences of everyday life. It is natural that we should try to impose concepts from our everyday world, like 'wave' and 'particle' on the subatomic world; but in fact all we know about the subatomic world is that if we poke it in a certain way then we get a certain response. 'What I do try to discredit', says Davies at the start of this essay, 'is what might be called naïve realism.' And he concludes, 'the concept of a particle is purely an idealized model of some utility in flat space quantum field theory.' The particle concept is, he says, 'nebulous' and ideally 'it should be abandoned completely.'

The snag is that we have nothing better, as yet, to replace it with. But it is with that cautionary word from the forefront of modern research, rather than any comforting undergraduate Linus-blanket of ideas about real little lumps being guided by probability waves, that I want to take you now into the particle world itself. Over the past fifty years, physics has revealed a wonderland of a subatomic world, populated by all kinds of strange objects. We call those objects particles, for want of a better name. What they really are, we do not know. The best theories we have explain the results of past experiments in terms of the interactions between these mythical beasts, and they predict the results of new experiments in terms of how the 'particles' interact with one another. By observing the world of high energy interactions, physicists work out rules which describe those interactions, and then they use the rules to predict the outcome of the next experiment. Good theories are the ones that get those predictions right; the best theories enable us to 'get right' the calculation of how the Universe came into being and then exploded into its present form. But that doesn't mean that they convey ultimate truth, or that there 'really are' little hard particles rattling around against each other inside the atom. Such truth as there is in any of this work lies in the mathematics; the particle concept is simply a crutch ordinary mortals can use to help them towards an understanding of the mathematical laws. And what those mathematical laws describe are fields of force, spacetime curved and recurved back upon itself in fantastic complexity, and a 'reality' that fades away into a froth of virtual particles and quantum uncertainty when you try to peer at it closely.

The concepts of particles and waves are the best we have, and the most straightforward way to describe the great advances in modern physics, in our understanding of the Universe, is in terms of particles. But they are only metaphors for something that we cannot properly comprehend or understand, and I apologize in advance for the necessity of using them. I feel like a blind man trying to explain the concept of colour to another blind person, having worked out a theory of colour based on touch. The surprise is not that our theories are flawed, but that they work at all.

Chapter Two

Particles and Fields

At the beginning of the 1930s, apart from the mystery of exactly how you should interpret the probabilities and uncertainty built into quantum theory, it looked as if physicists had a pretty good grasp of what the world is made of. There were four particles (the electron, the proton, the neutron and the photon), and there were two fundamental forces (gravity and electromagnetism), each of which had been known for a long time. Protons and neutrons together made up the nuclei at the hearts of atoms, and electrons occupied a more spread-out volume around the nucleus. Because protons carry positive charge, while electrons carry negative charge, each atom is electrically neutral, and the quantum physical arrangement of the electrons in different energy states, forced on them by their exclusive nature, gave each atom its unique chemical properties.

The power of the quantum theory was made impressively

clear when physicists looked at the size of atoms in terms of the uncertainty principle. An atom is about one hundred-millionth (10^{-8}) of a centimetre across, and the mass of an electron is just over 9×10^{-28}gm. The energy that an electron possesses when it is bound up in an atom can be calculated, using Bohr's early version of quantum physics, by treating it like a particle in orbit around the nucleus. Basically, this is the same math that can be used to calculate the energy of the planets in their orbits around the Sun. Such naive calculations give you a velocity for a typical electron 'in its orbit' of about 10^8 cm per second. Together, these figures give an approximate momentum for a typical electron in an atom, about 10^{-20}gm cm per second, or perhaps a little less. If the electron had much more momentum (and therefore more energy) than this, it would escape from the atom, because the electric forces between the electron and the nucleus would not be strong enough to hold on to such an energetic electron. So the *uncertainty* in the momentum, Δp, is itself something less than 10^{-20}gm cm per second. Multiply this by the uncertainty in the position of the electron, a bit less than 10^{-8}cm, and you end up with a $\Delta p \, \Delta x$ of no more than a few times 10^{-27}, very close to the value of $h/2\pi$. It is Heisenberg's uncertainty relation that determines the smallest size an atom can be. If the atoms were smaller, then there would be less uncertainty in the positions of the electrons, so there would be more uncertainty in their momentum, and therefore in their energy, and some would have enough energy to escape from the atom altogether, just as alpha particles 'tunnel' out of the nucleus. Quantum physics can explain, or predict, the size of atoms themselves, something which had to be accepted as being the way it was 'just because' it was the way it was, before quantum theory came along.

Protons and neutrons have much more mass than electrons, so they can have a larger uncertainty in momentum ($m \times v$) even though their velocities are smaller. Because their momenta are larger, they can be confined in a smaller volume (Δx) and still keep the product of uncertainties ($\Delta p \, \Delta x$) bigger than h cross. So nuclei are much smaller than the electron clouds which surround them, and their correct sizes are, once again, predicted by quantum uncertainty.

So, at the start of the 1930s, nature looked simple, and it seemed that physics had found the ultimate building blocks used by nature. Within a few years, however, the world began to seem a much more complex place, and within twenty years physicists had identified as many 'elementary particles' as there are different chemical elements. It took a revolutionary new approach to bring some order to this proliferation of particles, and to explain them in terms of a few units that are more basic still. Just as an atomic nucleus is imagined to be composed of protons and neutrons, so, for those who like to think in terms of particles, protons and neutrons themselves (and other particles) are now thought of as composed of quarks. But the very concept of a particle has itself undergone a dramatic change over the past fifty years. Just as photons are regarded as a manifestation of the electromagnetic field, so electrons (and other particles) can be thought of as manifestations of their own fields. Instead of a variety of fields and particles, interacting with one another, the Universe can be thought of as made up of a variety of interacting fields alone, with the particles representing the quanta of each field, manifested in obedience to the rules of wave-particle duality and the uncertainty principle. So before we look at those new developments, and how the physicists' view of our world has changed in the past fifty years, it seems appropriate to check out just what we mean by the concept of a field in physics.

Field Theory

The basic ideal of a field as the means by which an electric force is transmitted goes back to the pioneer Michael Faraday, who was born in Newington, in England, in 1791. Faraday's career was so remarkable that it is worth a little digression to sketch out just how he came to be one of the great scientists, and great popularizers, of the Victorian era.

His father was a poor blacksmith, and Michael Faraday himself received only the basic education available to the poor in those days, leaving school at the age of thirteen and becoming apprenticed to a bookbinder. But he had at least learned to read, and he had a voracious thirst for knowledge. So he read the books he was supposed to be binding, and at the age of fourteen became fascinated by an article in the *Encyclopaedia Britannica*. He read avidly about electricity and chemistry, and carried out his own experiments, as far as his limited resources permitted. In 1810, Faraday joined the City Philosophical Society, attending lectures on physics and chemistry in his spare time, and in 1812, at the age of twenty-one, his life was changed when he attended a series of lectures at the Royal Institution, given by Humphry Davy, a great chemist and the inventor of the safety lamp used by miners before the availability of electricity.

Faraday was enthralled by Davy's lectures and made extensive notes on them, which he bound up himself to keep as a permanent record.[1] He wanted desperately to become a full-time scientist himself, and wrote to the President of the Royal Society asking for advice and help, but did not receive a reply. As his apprenticeship finished in 1812, Faraday resigned himself to a life as a bookbinder. But he was rescued from this fate by an accident in which Davy was temporarily blinded by a chemical explosion. He asked the eager student, Faraday, to act as his assistant until he recovered his sight. Faraday performed satisfactorily, and when Davy recovered Faraday sent him the bound copy of his lecture notes. Davy was so impressed that when, a few months later in 1813, he needed an assistant at the Royal Institution, he offered Faraday the post. Faraday leapt at the opportunity — even at the modest salary of a guinea a week, less than he had been earning as a bookbinder.

Faraday spent the rest of his career at the Royal Institution, becoming Director of the Laboratory in 1825 and Professor of Chemistry there in 1833. He was a great experimenter and explainer rather than a mathematician, and was a very successful and genuinely popular lecturer, who founded the Royal Institution Christmas lectures for children, which continue to this day. By the time he died, in 1867, he had become a Fellow of

---

[1] You can still see them, in the Faraday museum at the Royal Institution in London.

the Royal Society and was widely recognized as one of the scientific giants of his day. But he was also modest, and along the way turned down the offer of a knighthood and *twice* refused the offer of the Presidency of the Royal Society. And, in his attempts to find a way to describe what happened when electric and magnetic forces act upon one another, he came up with the idea — what we would now call a model — of a 'line of force', which Maxwell then elaborated into the first field theory.

The idea can be simply understood in terms of the forces acting between electric charges. Like charges repel each other (positive repels positive; negative repels negative), while opposite charges attract. Faraday's lines of force could be thought of as mathematical lines stretching out from each charged particle in the Universe. Every line starts on a particle with one flavour of charge and ends on a particle with the opposite charge. Like stretched rubber bands, they tend to pull opposite charges together; but like squeezed elastic, squashed-up bunches of lines of force keep similar charges apart. The concept was enormously useful in getting a picture of what was going on, and the lines of force appear to have some real physical significance, because a tiny 'test' particle with a small positive charge, placed between two oppositely charged, static, large objects will drift from the positive to the negative object along a line of least resistance which (ignoring other forces such as gravity) will be along one of Faraday's lines of force.

All of the lines of force filling space constitute the electric field. In the classical field theory of Faraday and Maxwell, the field is continuous — there are no 'gaps' between lines of force and no breaks in the lines themselves. So the analogy becomes, more appropriately, with an elastic medium, filling the Universe, through which electric and magnetic forces are transmitted. This was the concept of the 'ether'. It seemed natural to Victorian scientists, versed in the nature of mechanical objects and the triumphs of engineering, but following the development of relativity theory and then quantum theory in the twentieth century this mechanistic view of the Universe was abandoned. The field is the field, not an elastic solid, and we cannot understand it fully in everyday terms. The most widely used 'explanation' of what the field is brings the point home, if you stop to think about it.

Before field theory was developed, it looked as if objects such as charged particles or magnets affected each other by an

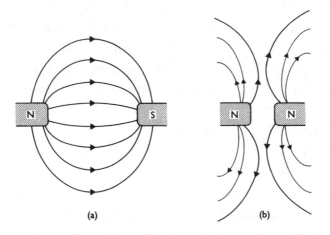

(a)   (b)

Figure 2.1 The concept of lines of force, invented by Michael Faraday, is useful in getting an image of how like magnetic poles repel one another and opposite poles attract.

action which reached out across the gap between them — action at a distance. Field theory says instead that the action is a local phenomenon, with each charged particle interacting with the field, and the field interacting with each particle, in a way which depends on all the other interactions between the field and its particles. The often-used analogy is with a spring. If you pull on either end of the spring, it stretches; if you squeeze it, it shrinks. The field, we are told, is like that. It can be stretched and compressed, and it links two particles in the way the matter in the spring joins its two ends together. The analogy seems homely and sensible enough. But what *is* the matter in the spring? It is a collection of atoms. And how do atoms interact with one another? Chiefly by electromagnetic forces. When the spring is stretched, the atoms move further apart; when it is squeezed, they are pushed closer together. What the analogy says is that the stretching and squeezing of the electromagnetic field is like pulling atoms apart or pushing them together — in other words, it is like stretching or squeezing the electromagnetic field! Perhaps, after all, it is best to stick with the equations that describe how particles and fields interact, rather than trying too seriously to get a mental image of what is going on.

Those equations are, in this case, Maxwell's equations. Maxwell gave us the first fully worked out field theory, one which applies to both electricity and magnetism, and which says that magnetism is equivalent to moving (dynamic) electricity. It is the classical (meaning non-quantum) theory of electrodynamics. Einstein's General Theory of Relativity describes the other force familiar from everyday life, and is another classical (in the same sense) field theory, this time of gravity. On the old, classical view of the world there were two basic components, material objects and the fields which linked them.

But the field is now the ultimate, fundamental concept in physics, because quantum physics tells us that particles (material objects) are themselves manifestations of fields. One of the first great surprises of quantum physics was the realization that a particle, such as an electron, had to be treated like a wave. In this first application of quantum principles, we learn to treat these matter waves as fields, with one field corresponding to each type of particle. For example, there is a general matter field, filling the Universe, which is described by the wave equations of an electron.[2] But, as the discovery that electromagnetic waves must also be regarded in particle terms showed, a field can be directly responsible for the existence of particles. Indeed, in the quantum world a field *must* give rise to particles. Quantum physics say that the energy in the field cannot be smoothly changing from place to place, continuously, as in the classical picture. Energy comes in definite lumps, called quanta, and every matter field must have its own quanta, each with a definite amount of energy, or mass. The particles are energetic bits of the field, confined to a certain region by the uncertainty principle. A photon is a quantum of the electromagnetic field; in the same way, by applying quantum principles a second time to the matter field electrons, we find that we recover the idea of the electron as a particle, as

---

[2] Strictly speaking, the wave function describes electron-positron pairs; more of this shortly.

the quantum of the electron matter field. This interpretation of particles as 'field quanta' is known as second quantization. It tells us that there is nothing else in the Universe except quantum fields. So the more we know about quantum fields, the better we will understand the Universe.

Fields come in different varieties. For example, some have an inbuilt sense of direction — they are called vector fields — while others do not. A field without an inbuilt sense of direction is called a scalar field. An example might be a field representing the temperature at every point in the room. Obviously, the field fills the room. If we place a thermometer at any place in the room it will register a temperature — less by the door, where there is a draught of cold air blowing in, and more by the radiator. But the thermometer is not propelled, by its interaction with the field, either towards or away from the radiator. The field has magnitude but not direction. The electric field, on the other hand, is a vector field. I can measure the strength of the field at any place, and also its direction. A little positively charged object dropped into the field will try to move 'along a line of force' away from the nearest positive pole of the field, and towards a negative pole.[3]

There is another important distinction which applies to the quantum fields. Although I have so far talked about electrons

---

[3] At a slightly more subtle level, even scalar fields can be associated with directions, because the field changes from place to place and objects interacting with the field seek a state of minimum energy. A ball falls downward in the Earth's gravitational field for this reason. But a tiny charged object will move along a line of force even in a perfectly uniform electric field, and this is the crucial distinction.

and photons and their related fields in more or less the same way, there is a fundamental difference between them. Electrons are members of the family of fermions, all of which have a spin which is half-integral, and fermions are not created or destroyed in the Universe today, except in matter-antimatter pairs. Photons, on the other hand, are bosons, and all bosons have zero or integer spin, and can be created and destroyed, and are nonexclusive. So there are two fundamentally different kinds of field in nature, one fermionic and the other bosonic. This, it seems, is what leads to the distinction between what we used to think of as particles and what we used to think of as forces.

When two particles interact, they do so, on the old picture, because there is a force between them. This force can be expressed in terms of a field, and that field can in turn be expressed in terms of particles by means of the second quantization. When two electrons come close to each other, and are repelled from one another, it is because, on the new picture, one or more photons have been exchanged between them. The energetic photon is a manifestation of the electric field around one or other of the electrons. It borrows energy from the uncertainty principle, pops into existence, whizzes across to the second electron and deflects it, before it disappears again. The first electron recoils as the photon leaves it, and the result is that the electrons are repelled from one another.[4]

---

[4] The analogy works so well for repulsion that I can't resist it. It is useless, though, at explaining how particles with the opposite kind of charge, such as an electron and a proton, *attract* one another. Unfortunately, the quantum world cannot always be described in terms of cosy analogies.

One kind of field, corresponding to fermions, produces the material world; the other kind, corresponding to bosons, produces the interactions which hold the material world together, and sometimes break bits of it apart.

The electromagnetic field around an electron can create virtual photons, provided that they are short-lived and do not stray far from home. The rule of thumb, from the uncertainty principle, is that such a virtual photon can only move half its own wavelength away from the electron before it must turn back and be reabsorbed. Longer wavelength corresponds to less energy, so less energetic virtual photons stray further from the electron. The result is a quantum picture of the electron as a charged region embedded in a sea of virtual photons, which get more energetic the closer you approach the electron itself.

Virtual photons, and indeed ordinary photons, can also create electrons, as virtual electron-positron pairs (see below), provided that they too are short-lived and exist only within the confines set by the uncertainty principle. And those electrons have their own clouds of virtual photons, and so on *ad infini-*

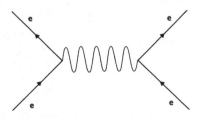

Figure 2.2   The classic Feynman diagram of particle interactions — in this case, two electrons interact by exchanging a photon.

tum. The froth of activity around an electron, or involved in the repulsion, or scattering, of one electron by another, is a far cry from the tranquil image most people associate with the word vacuum. But by applying the principles of quantum theory to the electromagnetic field in this way, physicists have been able to come up with a theory of quantum electrodynamics, and a living vacuum, which describes the interactions of electrons, photons and the electromagnetic field in quantum terms. The theory, known as QED, is one of the great triumphs of modern science, so successful at accounting for the electromagnetic interaction that it is regarded as the archetype of all quantum field theories and is used as the blueprint for constructing new theories to explain other interactions. But it suffers from one crucial flaw, related to the presence of that cloud of virtual particles around each electron.

Quantum physics tells us that an electron is surrounded by a cloud of virtual photons, and that any or all of these photons may be turning into electron-positron pairs, or into other pairs of virtual particles, before being reabsorbed by the electron. Extra energy is always being borrowed from the field, and from the uncertainty relation, literally without any limit to the complexity of the loops of virtual photons and virtual electron-positron pairs being produced. When the rules of quantum physics are applied with scrupulous care to calculate how much energy can be involved in these loops of virtual particles it turns out that there is no limit — the energy of the electron plus the cloud of virtual particles around it becomes infinite, and since the electron and its cloud are inseparable, at first sight that seems to mean that electrons must have infinite mass.

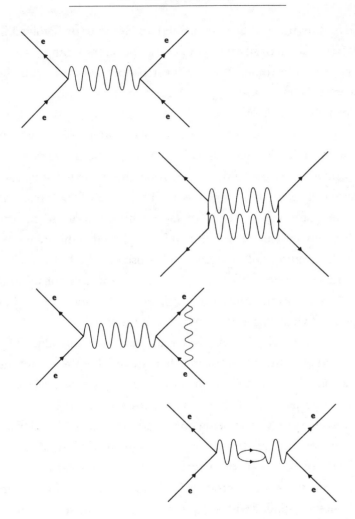

Figure 2.3 Quantum corrections to the laws of electrodynamics arise because real interactions are described by
more complicated diagrams in which closed loops represent the effects of virtual particles. These are the situations
that lead to infinities that can only be removed by the unsatisfactory trick of renormalization.

The way round this difficulty seems crazy, but it works. Mathematically, the infinite mass of the cloud around the electron is compensated for by assuming that a 'bare' electron (if such a thing could exist) would have infinite *negative* mass. With careful mathematical juggling, the two infinities can be made to cancel out, and to leave behind a mass corresponding to the mass we measure for an electron. The trick is called renormalization. It is unsatisfactory for two reasons. First, it involves, in effect, dividing both sides of a mathematical equation by infinity, which is something we were all taught at school (quite rightly) is forbidden. Secondly, even then it does not predict the 'correct' mass for the electron. Renormalization will give you a finite mass, but it could be *any* finite mass, and the physicists have to choose the right one and plug it in by hand. They can only solve the equations because they already know the answer they are looking for. The great thing is, though, that by putting in just one critical value by hand the equations then give the physicists very precise and accurate 'predictions' of the values of many other crucially important parameters — and that is why so many physicists have been happy to live with renormalization.

QED, including renormalization, works so well at explaining the behaviour of charged particles and electromagnetic fields within the framework of quantum physics that most physicists choose not to think too deeply about these problems. But if they could ever come up with a theory in which the infinities cancelled out by themselves, without the need for renormalization, then their joy would be unconfined. Quantum electrodynamics is the best and most complete quantum field theory we have, but it is still not perfect. The search for a perfect theory that will explain all the interactions in the Universe, and the existence of the Universe itself, is now beginning to bear fruit. But before we can taste the first fruits of that success, we have to come up to date by taking stock of the veritable menagerie of new 'particles' discovered in the past sixty years, and of the two new 'forces' needed to complete the count of quantum fields.

Two More Forces, and a Particle Zoo

At the beginning of the twentieth century, scientists knew of the existence of 92 different chemical elements. Each element was known to consist of its own variety of atoms, and the 92 different kinds of atoms were regarded as the fundamental building blocks of nature — though it did seem rather profligate of nature to require so many 'fundamental' building blocks. Thanks to the pioneering work of the Siberian Dmitri Mendeleev, who lived from 1834 to 1907, in the second half of the nineteenth century chemists had begun to appreciate the relationships between atoms with different weights. Mendeleev showed that when the elements were listed in order of increasing atomic weight, starting with hydrogen, then elements with similar chemical properties recurred at regular intervals throughout the resulting periodic table. This ranking of the elements by their chemical properties left some gaps in the table, and without knowing why or how the repeating pattern was produced the simple fact of its existence enabled Mendeleev to predict that new elements would be discovered to fill the gaps, and to predict what properties and atomic weights those elements would have. His predictions were subsequently confirmed, to the letter, by the discovery of 'new' elements.

When physicists became able to break the atom apart, and reveal its inner workings, they found that it contained three types of particle: electrons, protons and neutrons. Quantum physics, in the hands of Niels Bohr, was able to explain the observed properties of the chemical elements, and the structure inside the atom underlying the structure in Mendeleev's periodic table. The periodic table appeared first in a paper Mendeleev published in 1869; Bohr's explanation in terms of quantum physics came in the 1920s, less than sixty years later. But just as the nature of atoms was being understood in terms of subatomic particles, so the nature of subatomic particles began to look very far from clear.

The electromagnetic force, in its quantum form, was entirely adequate to explain the behaviour of negatively charged electrons as the partners, in an electrically neutral atom, to protons in the positively charged nucleus. But how could several protons, each with a positive charge, cluster together in the nucleus itself without being forced apart by electric repulsion? Rutherford, born in 1871, had established in 1910 that all of the positive charge in an atom is concentrated in a tiny nucleus. He also surmised, in the early 1920s, that there must be a neutral counterpart to the proton, a counterpart which he called the neutron, having about the same mass as the proton but no electric charge.

The presence of neutrons was necessary to explain why some atoms had very similar chemical properties to one another, but different weights. Chemistry depends on the number of electrons surrounding an atom, and that is always the same as the number of protons in the nucleus. So to change the weight of an atom without changing its chemistry you add, or subtract, electrically neutral particles (neutrons) to or from the nucleus. The atoms with the same chemistry but different atomic weights are called isotopes of the element concerned. James Chadwick confirmed that neutrons exist in a series of experiments in 1932, and received the Nobel Prize for his work in 1935.

That brief span of three years from Chadwick's discovery of the neutron to his receipt of the Nobel Prize marks the time when subatomic physics seemed simple, and it looked as if physicists only had four types of fundamental particle to worry about. The presence of neutrons even helped to explain the stability of the nucleus, since the positively charged protons could, to some extent, hide from one another behind the neutrons. But this help was not enough to explain the stability of nuclei, and with that realization the simplicity of the particle world began to disappear.

The first blow at the foundations came from a Japanese researcher, Hideki Yukawa. Yukawa had been born in 1907 (he died only in 1981), and after attending the universities of Kyoto and Osaka, in 1935 he was working for his Ph.D. (which he obtained in 1938) and teaching at Osaka University. In 1939, he returned to Kyoto as professor of physics. Like other physicists, Yukawa was puzzled about how the atomic nucleus held together. He reasoned that there must be another force, stronger than the electromagnetic force, which kept the protons in its grip even though the electric repulsion 'wanted' to separate them. But we don't see any evidence of such a strong force in the everyday world, so it must be a kind of force new to our experience, a force which only operates over a very short range, holding protons and neutrons together in the nucleus but permitting the individual particles (or, as we have seen, alpha particles) to fly free once they get beyond its range. Yukawa used an analogy with the electromagnetic force to describe his new force.

In electromagnetic field theory, the force results from the exchange of particles, virtual photons. Because photons have zero mass, the amount of energy a photon carries can be made vanishingly small by giving it a very long wavelength. So there is no limit, in principle, to the range over which electric forces can be felt — a virtual photon associated with an electron can interact, albeit very weakly and with very low energy, with another electron anywhere in the Universe — although, of course, the interaction is much stronger if the electrons are close together.

But what if the photons had mass? In that case, there would be a certain minimum amount of energy that would be required in order to make a virtual photon, ΔE. And the finite size of that packet of energy would set a firm time limit, Δt, on the life of such a particle, in line with Heisenberg's uncertainty relation. Since nothing can travel faster than light, this finite lifetime would mean that any such particle, in effect a 'massive photon', would have only a very limited range, since it has to return to its origin, or find another particle to bury itself in, before its allowed lifetime is up. So Yukawa reasoned, in 1935, that there must be another field, analogous to the electromagnetic field, associated with protons and neutrons. This field produces quanta which are (like photons) bosons, but which (unlike photons) have mass. And these bosons can only be exchanged by particles which 'feel' the strong field. It just happens, said Yukawa in effect, that electrons ignore the strong force.

The beauty of Yukawa's hypothesis was that it was possible to calculate just what the mass of this new kind of boson ought to be. Its range had to be not much more than the size of an atomic nucleus, or it would prevent alpha particles escaping, even with the aid of uncertainty, and produce other observable effects that are not, in fact, observed. And the size of a typical nucleus is only 10^{-12} cm, as refinements on Rutherford's team's pioneering scattering experiments show. From this single measurement and the uncertainty relation, Yukawa calculated that the carriers of the strong force must be particles with mass of about 140 MeV, more than 200 times the mass of an electron, but still only one-seventh the mass of a proton.

How could Yukawa's idea be tested? At that time physicists had no way of looking inside the nucleus to find the new bosons. But while the strong force was supposed to depend on the exchange of *virtual* bosons,[5] there was nothing in the

---

[5] It is because they are virtual particles, not real ones, that the ones 'inside' the nucleus can be involved in gluing the protons and neutrons together without contributing to the mass of the nucleus. In a proper quantum field theory treatment of the strong interaction, the same kind of problems with infinities arise as in QED, but they are also dealt with in the same way, by renormalization, as we shall see. But that involves a 'generation' of particles one step more 'fundamental' than the ones I am discussing now.

equations to prevent the equivalent real particles being produced anywhere where there was enough energy to do the trick. It happens that, according to the equations, these particles ought to be unstable. The mass energy locked up in them can be converted into other, stable forms. But they can also be made by the energy available in collisions between fast-moving (that is, energetic) particles. Today, giant particle accelerators like those at CERN, in Geneva, are used by physicists to smash beams of electrons and protons into one another, and into stationary targets, to manufacture showers of short-lived particles. These particles are made out of the kinetic energy of the colliding particles, in line with the equation $E = mc^2$, or if you prefer, $m = E/c^2$.

This is an important point to appreciate. The 'new' particles are not pieces of the particles that are being smashed together, broken off by the impact, but genuinely new, freshly manufactured out of pure energy. So the collisions can easily produce new particles which have a bigger rest mass than the particles involved in the collision, provided that the energy of motion involved is bigger than the required rest mass.[6]

---

[6] I was initially uncomfortable about using this terminology that takes no account of the fact that particles are 'really' field quanta. But 'particle' and 'field' are just labels we use for convenience. When I discussed how best to use these labels in my exposition with particle physicist Frank Close, he said it doesn't matter. When he looks at 'particle' experiments like those carried out at CERN, and tries to understand what is going on, he thinks in terms of the flow, or transfer, of momentum through the interactions being studied. That, he says, is what really matters. Both particles and field quanta have momentum, and as long as you know where the momentum is going the labels are of secondary importance.

In the 1930s, the only source of the necessary energetic particles was the Universe itself, which bombards the atmosphere of the Earth with the high-energy protons, electrons and (we now know) other particles that are collectively called cosmic rays. A cosmic ray, colliding with another particle high in the Earth's atmosphere, can create new particles, including the bosons of the strong force. The first high-energy physicists were observers who found ways to monitor the passage of cosmic rays through their experiments — the 'rays' affect photographic film, can be made to trigger sparks in apparatus rather like a grown-up version of a Geiger counter, and can be monitored in other ways. Once you've caught the fleeting passage of a cosmic ray, or a handful of cosmic rays, and photographed its trace, you can work out if it carries electric charge from the way its path bends in a magnetic field, and even deduce its momentum (and hence, eventually, its mass) from the amount by which its path is deflected by magnetic fields.

In 1936, one of the high-energy physics pioneers, the American Carl Anderson, was studying cosmic ray tracks produced in detectors on the surface of the Earth, and found traces of a particle heavier than the electron but lighter than the proton. It looked as if Yukawa's carrier of the strong force had been found; the particle was dubbed the mu-meson, or muon for short. In fact, further studies soon showed that the muon was not the carrier of the strong force. It didn't have quite the right mass, and it didn't display sufficient eagerness to interact with atomic nuclei. But in 1947 another cosmic ray physicist, the Englishman Cecil Powell, found a short-lived boson with exactly the right properties, including a mass very close to the value predicted by Yukawa, and a great enthusiasm for reactions with nuclear particles. It was called the pi-meson, or pion. The muon, it turns out, is one of the things produced when a pion decays. Yukawa received the Nobel Prize in Physics in 1949, the first Japanese person to be honoured in this way, and Powell was awarded the Prize in 1950. As for Anderson, he had already received the Physics Prize in 1936, the year he found the muon. But that was for a quite different discovery, one which didn't just add one member to the particle zoo but, by implication, essentially doubled the number of inmates overnight.

Paul Dirac, a British physicist who was born in 1902, was one of the pivotal figures in the quantum revolution of the 1920s. He fused the first version of quantum mechanics, developed by Werner Heisenberg, with Einstein's special theory of relativity, introducing the idea of quantum spin for the electron (an idea promptly taken over into other particles) in the process; he developed a very complete mathematical description of quantum theory, and wrote an influential textbook on the subject, still used by students and researchers today; and he played a major part in the development of QED, although to the end of his life (in 1984) he remained deeply unhappy with the business of renormalization, which he felt did no more than paper over the cracks in a flawed theory. For all that, outside of the inner circles of physicists, Dirac's best known contribution to our understanding of the Universe is his prediction, in 1928, that particles of the material world have their counterparts in the form of antimatter, mirror-image particles.

Ironically, for a theorist who achieved so much so accurately by design, Dirac's prediction of antimatter came about almost by accident, and at first he presented it to the world in an imprecise form. Dirac found that the equations he was working with, which described the behaviour of electrons, actually had two sets of solutions, not one. Anyone who has come across simple quadratic equations, ones involving the *square* of an unknown quantity, can soon get a grasp of why this was so. Squares are always positive. If you multiply 2×2 you get 4, and if you multiply -2×-2 you still get 4. So the 'answer' to the question 'What is the square root of four?' is either 2, or -2. Both answers are correct. The equations Dirac was dealing with were a little more complicated than this, but the principle was the same. They had two sets of solutions, one corresponding to the electron (which carries negative charge) and one corresponding to an unknown, positively charged particle.

In 1928, physicists only knew of two particles, the electron and the proton, although they strongly suspected the existence of the neutron. So Dirac's first idea was that the positively charged solution to his equations must represent the proton. It is a sign of the way even so great a physicist as Dirac was groping in the dark in the early 1930s that he saw no reason why the particles representing the positive and negative solutions should have the same mass. It is only now, with hindsight, that we can say 'of course' the electron's counterpart must have the same mass as the electron, and that the proton is too heavy to be its companion. Initially, scarcely anyone seems to have taken Dirac's proposal seriously —

there was certainly no concerted hunt for the hypothetical new particle, as there would be if a similar hypothesis were made today. Physicists dismissed the idea that Dirac's calculations were telling them anything significant about the world. The existence of the other solution to the equations was ignored, just as an engineer working with quadratic equations will ignore one of the solutions to those equations and keep the one which obviously applies to the job in hand, building a bridge or whatever it might be.

But in 1932 Anderson was studying cosmic rays, using a cloud chamber, a device in which the cosmic ray particles leave trails behind them, like the condensation trails produced by high-flying aircraft. These trails are photographed, and the patterns they make can then be analysed at leisure. One of the things Anderson did was to investigate how the trails changed under the influence of a magnetic field, and he found some trails that bent by exactly the same amount as the trail of an electron but in the opposite direction.[7] This could only mean that the particles responsible had the same mass as an electron, but the opposite (positive) charge. The new particles — called anti-electrons or, more commonly, positrons — were

---

[7] The trail left by an electron moving clockwise in a magnetic field is, of course, identical to the trail made by an equivalent positron moving anticlockwise. Anderson's achievement lay as much as anything in the way he determined which direction the particles making their trails were moving in, and such subtleties help to explain why his work merited the award of a Nobel Prize. It is easy enough for me to say 'Anderson measured the curvature of particle trails and found positrons'; it was much harder for him to do the work.

soon identified with the particles predicted by Dirac's equations, and this was the work which earned Anderson his Nobel Prize. Dirac received the Prize, jointly with Erwin Schrödinger, in 1933.

The positron was discovered in the same year as the neutron although, in fact, the evidence for these positively charged particles had been around in cosmic ray tracks for some time, but had been mistaken for tracks of electrons moving the opposite way. Extending Dirac's calculations to all the atomic particles, this gave physicists six (plus the photon) to worry about — the electron and positron, the proton and a (presumed) negatively charged antiproton, and the neutron and a (presumed) antineutron.[8] The laws of physics require that when a particle meets its antiparticle counterpart the two annihilate in a burst of energetic photons (gamma rays). The positron and electron cancel each other out, as far as the material world is concerned. In the same way, running the equations in the other direction, if enough energy is available electron-positron pairs, or other particle-antiparticle pairs, can be created. But you must always pair up each particle with its precise mirror image in these interactions — not, for example, proton with antineutron. All of these predictions were borne out by experiments, although the antiproton and antineutron were

---

[8] Even though the neutron carries no electric charge, and therefore the antineutron carries no charge, they are as distinct as the members of the other two pairs, and the implications of Dirac's calculations are as profound for them as for the others. The photon is in effect its own antiparticle, a subtlety I'll discuss later.

not detected until the middle of the 1950s. And this interrelation between matter and energy, always obeying $E = mc^2$ as well as the rules of quantum physics, is, as we have seen, fundamental.

The positron and neutron were discovered in the same year, 1932. The muon was discovered in 1936, the pion in 1946. By then, it was clear that matter came in two varieties: some particles which feel the strong force (protons and neutrons, and the pions which carry the force) and some which don't (the electron and, it turns out, the muon).[9] This led to a new way of classifying particles, both the 'material' particles and the force carriers. Things which feel the strong force are called hadrons, while things which don't feel the strong force are called leptons. All leptons are fermions, and have half-integer spin. The only leptons we have met so far are the electron and the muon, which is identical to the electron except for its much greater mass. The hadrons that are also fermions ('matter') are called baryons. Protons and neutrons are baryons. The bosons that carry the forces between particles are now known specifically as mesons. The pion is a meson, and comes in three varieties. There is a neutral pion, which has no charge. When a proton and a neutron exchange a neutral pion, they are held together but remain unchanged. Protons also exchange neutral pions with each other, and so do neutrons. And there are two charged pions, positive and neg-

---

[9] Strictly speaking, I should include their antiparticles in this description. But everything that applies to protons, say, also applies to antiprotons, and so on.

ative, which are the antiparticles of each other.[10] If a proton gives a positively charged pion to a neutron, the proton becomes a neutron and the neutron becomes a proton. This is exactly the same as if the neutron gave a negative pion to the proton. Every variation on the exchange also helps to hold the proton and neutron together.

Already the number of particles needed to describe even the atom is growing. But there is still one more particle, and one more field, to add to the list.

Back at the end of the nineteenth century, Rutherford, working first in Cambridge and then in Canada, had discovered that uranium emits two kinds of radiation, and investigated their properties (he also discovered a third form of radiation, gamma rays, later identified with energetic photons). One of these 'rays', alpha radiation, was later discovered to be made up of helium nuclei, two protons and two neutrons bound together in a stable state. The other, which he called beta radiation, was later shown to be identical to electrons. So atoms can eject electrons. These electrons do not come from the cloud surrounding an atomic nucleus. Rutherford and his colleague Frederick Soddy were able to show, early in the twentieth century, that when a radioactive atom emits an electron it becomes an atom of a different element. Later studies

---

[10] The antiparticle equivalent of the neutral pion turns out to be indistinguishable from the original; as far as any observable interaction is concerned, the neutral pion is its own antiparticle. The same is true of the photon; although in principle you can set up equations that describe antiphotons, in practice the photon and antiphoton are the same.

showed that a neutron in the nucleus is converted into a proton, while the electron is ejected, producing a new nucleus corresponding to an atom of a different element. In fact, this process happens only in a few, unstable nuclei. Most neutrons, in most atoms, are quite happy as they are. But a neutron that is isolated, away from an atomic nucleus, will decay, as it is called, into an electron and a proton in just a few minutes. The process is called beta decay, and it must involve both another force and another particle, in addition to the ones mentioned so far.

Historically, it was the extra particle that physicists — or, rather, one physicist — came up with first. Beta decay was a main topic of research among physicists in the first decades of the twentieth century, and among their more surprising discoveries the physicists found that the electrons produced in the decay could emerge with various amounts of energy. The proton and electron, produced when a neutron decays, together have a total mass about 1.5 electron masses less than the mass of the neutron.[11] So this much energy ought to be available, shared between the proton and electron, as kinetic energy. When the proton is left in an atomic nucleus, of

---

[11] It is worth pointing out that there is no way in which an electron can be thought of as existing 'inside' a neutron — that a neutron might, on such a picture, be a composite of an electron and a proton, held together by electromagnetic forces. The uncertainty principle doesn't permit an electron to be confined within the diameter of even an atomic nucleus, let alone within a single neutron. To turn a neutron into a proton and an electron you *must* invoke, among other things, the mass-energy equation of relativity, which allows mass to be converted into energy and thence into another form of mass. Each electron produced by a beta decay is a newly created particle.

course, it doesn't move much, so it seemed that almost all of the extra energy must go to the electron, giving it kinetic energy in addition to its rest mass, and that every electron produced in this way by a radioactive atom ought to run off with a large, and predictable, amount of kinetic energy. But experiments showed that the actual energy of a beta decay electron is always less than the energy available, and sometimes a lot less. Where had the extra energy gone?

Wolfgang Pauli, a physicist who had been born in Austria in 1900, came up with the answer in 1930. There must be *another* particle produced, as well as the electron and proton, and this extra particle was running off, unseen, with the 'missing' energy. The particle required to do the job must have zero mass and no electric charge, or it would have been noticed by the experimenters.

Such a bizarre possibility did not meet with instant acclaim among physicists. It seemed too easy, and held out the threat

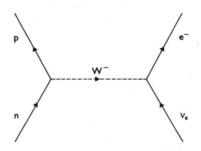

Figure 2.4 All fundamental interactions can be represented by Feynman diagrams. In this case, the diagram shows beta decay at work at the level of neutrons and protons.

of invoking a new, undetectable particle to explain every puzzling phenomenon in experimental physics. But Pauli persisted in promoting the idea, and won support, in 1933, from an Italian-born physicist a year his junior, Enrico Fermi. Fermi took up Pauli's idea and put it on a respectable footing by introducing a new force into the calculations, the so-called 'weak' nuclear force.

Field theory required a new force to account for beta decay anyway. It couldn't be the strong force that was responsible (electrons don't 'notice' the strong force), and it certainly wasn't electromagnetism or gravity. Fermi modelled his theory as closely as he could on QED, and came up with the idea that when a neutron changes into a proton it emits a particle which is the carrier of the new field, a charged boson usually written as W^-. The boson (now called an intermediate vector boson) carries off the electric charge and excess energy, while the neutron changes into a proton and recoils. But this boson is very massive (just how massive the early, incomplete version of the weak theory could not say).[12] It doesn't only carry the energy needed to make the electron; it has an enormous content of virtual energy, borrowed from the vacuum, so it is very unstable and doesn't live very long at all. Indeed, it doesn't live long enough to interact with any other particles, but quickly

---

[12] In fact, Fermi's original version of the model had all of the interactions occurring at a point, in effect giving the W particle zero range and infinite mass. The idea of using a particle with finite mass to describe the weak force dates from 1938, when it was introduced by the Swedish physicist Oscar Klein.

gives back its borrowed energy to the vacuum, allowing the rest to form into an electron and the new particle in the same way that an energetic photon can disintegrate into an electron and a positron. The electron is a lepton, so, strictly speaking, in order to conserve the total number of leptons in the Universe the extra particle that is produced must be an *anti*lepton. (Since we start with one baryon, a neutron, and end with one baryon, a proton, baryon 'number', as it is called, is also conserved.) Fermi called the extra particle a neutrino, meaning 'little neutral one'; today we would call it an electron antineutrino.

In 1933, the English journal *Nature* rejected a paper from Fermi setting out these ideas, saying it was 'too speculative'. But his work was soon published in Italian, and not long after in English. Evidence for the existence of the neutrino came in 1953, from experiments making use of the flood of such particles produced by a nuclear reactor. It has all the properties (or lack of properties) expected from the theory, although there is some speculation today that neutrinos may actually have a very small mass, far less than the mass of an electron.

So in the early 1950s physics had enough particles and fields to explain how atoms behave. The weak field, and the interactions that it mediates, is crucially important for the processes of nuclear fusion and fission, the manufacture of the elements in the stars, the fact that the Sun is hot, and the power of the atomic bomb. Electromagnetism remained its familiar self, and gravity refused, as stubbornly as ever, to be brought in to the quantum fold. Not all fields will succumb to the trick of renormalization, and every attempt to tackle the problems with infinities that had been tamed[13] with QED failed in the context of gravity. Just two leptons (and their antiparticle counterparts) were known, the electron and muon, and each of them had associated with it its own type of neutrino. So attention focused on the particles that are governed by the strong force. But for a decade, the more physicists probed the nature of hadrons, the more confusing a picture they came up with.

In 1932, the material world could, it seemed, be explained in terms of three particles. In 1947, there were half a dozen (and their antiparticles). By the end of 1951, there were at least fifteen 'fundamental' particles, and the list was just beginning to grow. Today, there are more particles in the list than there are elements in the chemists' periodic table. The decade of the 1950s saw physicists manufacturing new kinds of hadrons every time they opened up new particle accelerators, creating more and more members of the particle zoo, not quite out of thin air, but out of pure energy. The energy came from machines, ever bigger and better machines, in which charged particles such as electrons and protons were accelerated by electromagnetic forces and smashed into one another, or into targets of solid matter — which means into atomic nuclei, since the accelerated particles brush through the electrons in clouds around atoms like a six-inch shell moving through sea mist. No material object can be accelerated to the speed of light, so as more and more energy went into these ex-

---

[13] Or, as Dirac would no doubt have said, swept under the carpet.

periments the particles did not go faster and faster. Once their speeds were up to a sizeable fraction of the speed of light, they increased in mass instead. And when they were involved in collisions — interactions — all this extra mass was available to create other particles (almost invariably short-lived) which showed up as tracks in bubble chambers and other detectors. Generally, of course, each new particle made in this way is accompanied by an antiparticle partner; both lepton number and baryon number are conserved, although mesons can be manufactured at will.

Once again, I should stress that there is no sense in which these 'new' particles could be thought of as being 'inside' the protons, or whatever particles were being used in the colliding beam experiments. The particles were being made out of the energy fed into the machines. The new particles were given names, and their family characteristics identified and labelled, sometimes with quixotic terms such as 'strangeness'. Particle physics was in a stage very similar to chemistry before Mendeleev, when the elements were identified and their properties determined and compared with one another, with no idea of how or why those properties and family relationships were produced. The step forward for chemistry came with the periodic table of elements, and its later interpretation in terms of the structure inside the atom. The step forward for particle physics came in the early 1960s, with the development of a 'periodic table' for the particles, and a few years later with the interpretation of this new periodic table in terms of the structure within hadrons themselves.

The Eightfold Way: Order Out of Chaos

Field theories were not making much progress towards an understanding of the multiplicity of hadrons at the end of the 1950s. There were problems with infinities, like the ones that have to be renormalized in QED, and there was also some difficulty with the need to invoke a separate field for each particle — OK, perhaps, when you have two or three fundamental particles, but more and more disquieting when the count rises to dozens, and then above a hundred. Most theorists abandoned field theory in the early 1960s, trying other approaches to the problem of the strong interaction. I won't discuss those approaches here, since in the 1970s it was field theory that triumphed. But although the impetus for finding patterns among the properties of hadrons came partly from ideas developed in the context of field theory in the 1950s, the 'periodic table' of the particles stood on its own merits, as a classification system like Mendeleev's table, at the start of the next phase of development of particle physics.

The classification system was arrived at independently by two physicists, the American Murray Gell-Mann (born in 1929) and the Israeli Yuval Ne'eman, born in 1925. Ne'eman's education and career were interrupted by the fighting in the Middle East, following World War Two, during which Israel emerged as a nation in the region that had previously been Palestine. Ne'eman stayed in the Israeli armed forces after this period of disturbance, but found opportunities to study as well as to carry out his military duties. Although his first degree was in engineering, Ne'eman's interests led him towards problems in fundamental physics during the 1950s, when he served as a Military Attaché at the Israeli Embassy in London, and worked for a Ph.D., which was awarded in 1962, at the University of London. Gell-Mann's career followed a more conventional route, from Yale University to MIT, where he earned his Ph.D. in 1951, and then to Princeton, the University of Chicago (where he worked for a time with Fermi), and, in 1955, Caltech. Gell-Mann was responsible for the idea of 'strangeness' as a quantifiable property of particles, an idea introduced into particle physics to account for some of the new phenomena being observed in high energy interactions in the early 1950s.

Strangeness is just a property that fundamental particles seem to have (or, rather, it is a property that we need to put into our models if we want to think of the world as being made of particles). It is no more, and no less, mysterious than electric charge. Some particles carry charge, some do not, and charge comes in two flavours, which we call + and −. If we are being more precise, and include zero charge, we have three choices, +1, 0 and −1. Strangeness varies from particle to particle, and there are more options than with charge, but the principle is the same. Strangeness can be 0, −1, +1, +2, or even bigger. And strangeness has to be conserved in strong particle interactions. Just as a neutron can only turn into a proton by an interaction which produces an electron to balance the electric charges (and an antineutrino to preserve lepton number unchanged), so strangeness has to balance, by the creation of particles with the appropriate amount of strange 'charge', during strong interactions. This restricts the number of interactions allowed, in line with the 'strange' results physicists were obtaining in the 1950s — hence the name.

By using rules of this kind, Gell-Mann and Ne'eman were each able to group the new particles, together with the old familiar ones, into patterns according to their charge, spin, strangeness, and other properties. Gell-Mann called this the 'eightfold way', in conscious tribute to the 'eight virtues' of Buddhist philosophy, because some of the patterns he found initially involved particles in groups of eight. In fact, the system includes families with 1, 8, 10 or 27 members, in which each member of a particular family represents a variation on some basic theme. The system was proposed in 1961, and in 1964 Gell-Mann and Ne'eman, together acting as editors, produced a book, *The Eightfold Way*,[14] in which their own original papers and other key contributions to the new understanding of the particle zoo were reprinted. By then, the classification system had made a triumphant prediction of the existence of a new particle, putting it on the same secure footing that Mendeleev's table had just before the development of quantum physics.

The pattern of the eightfold way, extended to group one family of baryons in a pattern of ten, had a gap in it. One particle was needed to complete the picture, and Gell-Mann called it the omega minus (Ω^-), after the last letter of the Greek alphabet. The gap in the pattern 'belonged' to a parti-

---

[14] Benjamin, New York.

cle with negative charge, a strangeness of −3 and a mass of 1680 MeV. Just such a particle was found, in 1963, by researchers following up the prediction at the Brookhaven Laboratory, in New York, and at CERN, in Geneva. It took sixty years for Mendeleev's table to be interpreted in terms of a complete theory of the structure of the atom. It took little more than ten years for the eightfold way to be interpreted in terms of a complete theory of the structure within hadrons, and it only took as long as that because many physicists were initially reluctant to accept the idea, put forward as early as 1964, by Gell-Mann and separately by George Zweig, that 'fundamental' particles such as protons and neutrons are actually made up of peculiar particles called quarks, which come in threes and which have, heretical though it may seem, charges which are a fraction of the charge on an electron.

Quarks

Looking back over a span of more than thirty years to the genesis of the quark model of matter, it is hard to tell just how seriously even the proponents of the model took it at first. The idea that protons and neutrons, and other particles, were actually made up of triplets of other particles, some with a charge of ⅓ of the electron's charge, some with a charge of ⅔, ran so much against the grain of everything that had been learned since the closing years of the nineteenth century that it could only be presented, at first, as a device, a mathematical trick for simplifying some of the calculations and giving an underlying structure to the patterns of the eightfold way. In itself, that was, and is, no bad thing. It reminds us that *all* of our models of fundamental particles and their interactions are no more than artificial aids to help us to get a picture of what is going on in terms that seem familiar, or at least recognizable, from everyday life.

But it is ironic that as the quark model has become increasingly well established, in recent years many accounts of particle physics seem to have lost sight of the fact that even the best of our models are no more than aids to the imagination, and their accounts have begun to present an image of protons, neutrons, and the rest as made up of 'real' little hard lumps, the quarks, which rattle around inside what we used to think of as the 'fundamental' particles. The image is beguilingly reminiscent of the earlier vision of the atom as being made up of little hard lumps — electrons, protons and neutrons — and it is just as inaccurate.

Whatever its basis in 'reality', however, the quark model very neatly explains the interactions of the particle world.[15] The everyday particles, the proton and neutron and the pions which carry the strong force, can all be described simply in terms of two quarks, which are given arbitrary labels in order to distinguish them from one another. One is called 'up' and the other is called 'down'. The names have no significance;

---

[15] There were rival ideas in the 1960s, of course. Indeed, right up to 1970, the line of attack I describe here constituted less than half of all the theoretical papers tackling the problems of high-energy physics published each year. Once again, in the interests of brevity I have stuck to what turned out to be the main trail, and even left out most of the backtracking, blind alleys and retracing of steps that took place with the development of the quark theory itself.

physicists could just as well, if they wish, call one quark 'Alice' and the other one 'Albert'. The up quark has a charge of $\frac{2}{3}$, and the down quark a charge of $-\frac{1}{3}$, on this picture, and a proton is made up of two up quarks and one down, giving a total charge of $+1$, while a neutron is made up of two down quarks and one up, giving a total charge of zero. Pions are then 'explained' as being formed of *pairs* of quarks, always with a quark and antiquark paired together. Up plus anti-down makes the pi$^+$, down plus anti-up make the pi$^-$ and up plus anti-up *and* down plus anti-down together make the pi^0.

All of this is, so far, no more than an *aide memoire*, a mnemonic for constructing the fundamental particles. But the power of the mnemonic became apparent when Gell-Mann and Zweig invoked a third quark, the 'strange' quark, to account for the property of strangeness. By successively replacing one, two or three of the quarks in ordinary matter with a strange quark, they could build up particles with strangeness number of -1, -2 or -3 (the negative numbers are a historical accident of the way strangeness is defined). The proton and neutron have zero strangeness, because they contain no strange quarks; the omega minus has strangeness -3 because it is built up from three strange quarks, and so on. The whole of the eightfold way pattern fell naturally, in this way, out of the possible combinations of triplets of quarks and of quark-antiquark pairs. By assigning a definite mass to each quark, with the strange quark being about 50 per cent heavier than the other two, it even gave the right masses for all the known particles. But did the quark model have any *physical* significance?

Even Gell-Mann, who coined the name 'quark' from a line

in *Finnegans Wake,*[16] was almost coy about the concept in the paper in which he introduced it. He said:

> It is fun to speculate about the way quarks would behave if they were physical particles of finite mass (instead of purely mathematical entities as they would be in the limit of infinite mass) . . . a search for stable quarks of charge $-\frac{1}{3}$ or $+\frac{2}{3}$ and/or stable diquarks of charge $-\frac{2}{3}$ or $+\frac{1}{3}$ or $+\frac{4}{3}$ at the highest energy accelerators would help to reassure us of the non-existence of real quarks.[17]

Did Gell-Mann himself really believe in the reality of quarks, but try to slide the concept into the physics literature as if it were just an amusing trick? Or was he as doubtful about the whole business as these words suggest? There is no doubt that Zweig took the idea seriously — and, equally, there is no doubt that he got precious little in the way of praise, and plenty in the way of brickbats, as a result.

George Zweig was born in Moscow, in 1937. But he moved to the United States with his parents as a baby, and obtained a B.Sc. in mathematics from the University of Michigan in 1959. He then moved to Caltech, to begin his career in research, and spent three years struggling with a high energy experiment at an accelerator called the Bevatron, before he

---

[16] 'Three quarks for Muster Mark,' which, from context, gives the pronunciation to rhyme with 'bark', not with 'pork'.

[17] From Gell-Mann's paper in *Physics Letters*, volume 8 page 214, 1964; also quoted by Andrew Pickering, in *Constructing Quarks*, Edinburgh University Press, 1984, page 88.

decided to concentrate on theory and began to investigate physicists' understanding of the nature of the material world, under the guidance of Richard Feynman. As a newcomer to the field, he perhaps lacked some of the caution, or tact, of his elders, and when he realized that the eightfold way patterns of mesons and baryons could be explained in terms of combinations of two or three sub-particles, he immediately treated these sub-particles as real entities, which he called aces, and described them as such in his work. This bull-at-a-gate approach seems to have filled his superiors (not including Feynman) with horror — a horror only compounded by the success of what they saw as a naïve, unrealistic approach. In 1963, Zweig went on a one-year fellowship to CERN, where he wrote up his work for publication in the form of CERN 'preprints', concluding 'In view of the extremely crude manner in which we have approached the problem, the results we have obtained seem somewhat miraculous.'[18] But were these really his own views? Or did someone else put the remarks about crudeness into his mouth — or pen? For the appearance of these reports at all in 1964 was itself something of a miracle. When the student Zweig presented his first drafts to his superiors at CERN they were dismissed out of hand, and he recalled in 1981, in a Caltech publication how:

> Getting the CERN report published in the form that I wanted was so difficult that I finally gave up trying. When

---

[18] The CERN preprints both appeared in 1964, and are numbered 8182/TH401 and 8419/TH412. A more accessible, but secondhand, account is given by Pickering.

the physics department of a leading university was considering an appointment for me, their senior theorist, one of the most respected spokesmen for all of theoretical physics, blocked the appointment at a faculty meeting by passionately arguing that the ace model was the work of a 'charlatan.'[19]

Whatever the theorists thought in 1964, the quark model continued to provide at the very least a handy rule of thumb for calculating how hadrons ought to behave. And with the latest generation of particle accelerators, the experimenters had the means to test the hypothesis by shooting electrons at protons, with so much energy that they ought to scatter off individual quarks inside the protons. The experiments that in effect 'X-rayed' the proton involved an accelerator two miles long, at Stanford in California (the Stanford Linear Accelerator, or SLAC), where electrons were accelerated to energies of more than 20 thousand million electron volts (GeV). The way the electrons scattered from the protons in the targets

---

[19] Isgur, page 439, 1981. In the same report, Zweig says that 'Murray Gell-Mann once told me that he sent his first quark paper to *Physics Letters* for publication because he was certain that *Physical Review Letters* would not publish it.' It is interesting that Gell-Mann received the Nobel Prize in 1969, specifically for his other contributions to particle physics, notably strangeness and the eightfold way. Even in 1969, quark theory was not the obvious way forward in understanding the particle world, and ranked low on the list of achievements cited. Zweig has not (yet) received the Prize, even though quark theory is now fundamental to our understanding of the Universe, and his 1964 version was much more completely worked out, in detail, than Gell-Mann's. Is Zweig still paying a price for his youthful presumption? If not, perhaps the Nobel Committee will one day wake up to their oversight.

they were directed at clearly implied that there were concentrated regions of electrically charged mass-energy inside the proton, just as Rutherford's experiments, all those years ago, had shown that there is a concentrated nucleus inside every atom. At about the same time, in the late 1960s, experiments at CERN, in which neutrino beams, instead of electron beams, were used to probe protons, showed that there must also be electrically neutral 'matter' inside the proton. But no matter how hard the protons were bombarded, and no matter what they were bombarded with, it proved impossible, as it has proved ever since, to knock one of the presumed quarks out of it.

The explanation of the neutral matter associated with quarks inside the hadrons was simple in principle, although it raised new questions about what kind of fundamental theory could be constructed to explain what was going on. Just as protons and neutrons are held together by the exchange of pions, the carriers of the strong force, so quarks must be held together in some way, by an exchange of particles which were dubbed 'gluons' because they glue the quarks together to make protons, neutrons and so on. At first sight, this looks as if it means that we have a fifth force to worry about. But current thinking suggests that the glue force is the real 'strong' force of nature, and that the so-called strong interaction of nuclear physics is actually a side-effect of the glue force, in a way that is crudely similar to the way in which residual traces of the electric forces which hold atoms together in molecules provide a kind of weak electromagnetic force between different molecules.

But alongside the experimental successes of the quark model in the second half of the 1960s, there were problems. Why did quarks only come in threes, or in quark-antiquark pairs? The most profound puzzle, ironically, concerned the omega minus, the crowning prediction of the eightfold way, and particles which shared one important property with it. The omega minus is viewed, on the quark model, as a particle composed of three strange quarks. But all of these quarks have to be spinning the same way, so they are in identical states. Similarly, the experimenters had found a type of particle which could best be explained as a set of three up quarks spinning the same way, and one which consisted of three down quarks, all with the same spin. Yet the quarks are fermions, and the Pauli exclusion principle says that no two fermions, let alone three, can be in the same state. Is it possible that quarks don't obey the exclusion principle? Or is there some way in which the three quarks inside the omega minus (and in other particles built of three 'identical' quarks) can be distinguished?

A good theory of quarks would have to answer all these questions and more besides. The 'good theory' that was needed turned out to be a field theory. But the revival of field theory in the 1970s, which led on to a good theory of quarks and now, in the 1980s, to hope of a unified theory of all the fields, came from a breakthrough in the study not of hadrons but of leptons and photons — a new theory that combined the electromagnetic and weak forces into one description, the electroweak theory. But before we can look at how that new theory was developed, and how it in turn helped theorists to find a better model of the strong force, we need to delve once more into the mathematicians' box of tricks, and find out how to apply one of their most useful conceptual tools.

Gauging the Nature of Things

The interactions of ordinary matter all involve, on the new picture, just four particles — the up and down quarks, the electron and its neutrino. When a neutron decays into a proton, emitting an electron and an antineutrino in the process, what quark theory says is that a down quark inside the neutron changes into an up quark and emits a W^-, which in turn produces the electron and antineutrino. Another way of looking at this kind of interaction is as an exchange in which a down quark gives a virtual W^- to a neutrino, converting it into an electron and converting itself into an up quark. The electron and its neutrino are the leptonic equivalents to the up and down quark in the hadron world. These interactions are all described schematically by scattering diagrams like the one shown in Figure 2.2; mathematically, a particle travelling forward in time is the same thing as its antiparticle equivalent travelling backwards in time, so one basic diagram can stand for any or all of the fundamental interactions.

Historically, of course, an understanding of the weak force began to be developed before the idea of quarks was mooted, so the equations and diagrams were, and often still are, expressed in terms of protons and neutrons rather than up and down quarks. It makes no difference to the general thrust of the argument, and I shall use both descriptions interchangeably. But it is worth remembering that at this level of description, dealing only with the common form of matter that makes up the Sun and stars, distant galaxies, interstellar matter, planets and ourselves, we are dealing with a limited number of fundamental particles, just four (the up and down quarks, the electron and its neutrino). Almost all of the physics described in this book so far, and the evolution of the Universe, would be the same if these were, in fact, the only four types of particle that existed.

When physicists attempted to construct a more complete field theory of the weak interaction, in the 1950s, they naturally looked to the field theories they already had — gravity and, especially, electromagnetism — to decide what sort of properties a 'good' theory ought to have. One of the most powerful concepts which can be used to describe these fields is the property of symmetry. The electric field is, for example, symmetric in terms of the forces between charged particles. If you were to lay out an array of charged particles, some positive and some negative, and to measure all of the forces acting between them, and then somehow it was possible to reverse the polarity of every charge, positive for negative and negative for positive, while keeping them all in place, then you would find that the forces acting on each particle were exactly the same as before. Such a symmetry is called a global symmetry — every charge (strictly speaking, every charge in the Universe) has to be reversed at the same time to retain the original field of force.

Other laws of physics, or properties of particles, can be described in terms of symmetries. Positive and negative charges

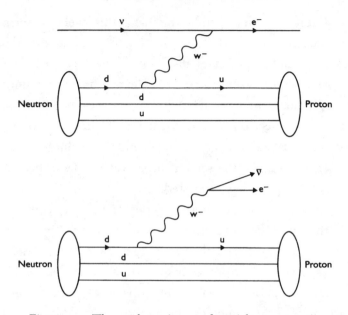

Figure 2.5 The modern picture of particle processes in-
terprets beta decay at a deeper level than the one shown
in Figure 2.4. Here, one of the down quarks inside a neutron emits a W⁻ particle, and becomes an up quark (so the
neutron becomes a proton). The released W⁻ particle may
interact with a passing neutrino, converting it into an electron (top picture); or, more likely, it will decay into an
electron and an antineutrino (bottom picture).

can be thought of as mirror-image, opposite versions of some
fundamental 'state' of things. If we ignore electromagnetic
forces, however, and look at the rest of the properties of the
proton and the neutron, they are very similar to one another.
So similar, indeed, that physicists regard them as two possible
'states' of a fundamental entity which they call the nucleon.

What decides whether a nucleon is a proton or a neutron (leaving aside, as I have said, the question of charge)? Just as the terms 'positive' and 'negative' are introduced to describe the different versions of charge, and just as quarks are arbitrarily given names such as 'up' and 'down', so physicists give a name to the property that distinguishes neutron from proton. They call it isotopic spin, and think of it as like an arrow, associated with each nucleon, that points either up and down or across. But it doesn't 'point' in the three-dimensional space of everyday life. Instead it is thought of as 'pointing' in some abstract, mathematical space which represents the internal structure of the nucleon.

We can imagine changing the isotopic spin of every nucleon simultaneously, so that every proton in the Universe became a neutron and every neutron became a proton. That would be equivalent to rotating the orientation of the isotopic spin of every nucleon by 90°, through a right angle. The point

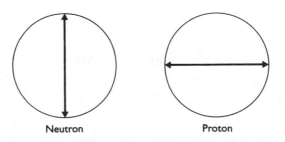

Neutron   Proton

Figure 2.6 The difference between a neutron and a proton can be represented by the direction in which a mythical 'internal arrow' attached to each nucleon is pointing. This arrow is dubbed isospin.

of all this is that the strong force is unaffected by such a transformation, just as the electric force is unaffected by reversing the sign of all electric charges. There is a fundamental symmetry between the two nucleon states, between the proton and the neutron — or, at a deeper level, between the up and down quarks. So when an individual neutron actually does change into a proton, the local symmetry, for that particular nucleon, is disturbed. There has been a local symmetry transformation. But the laws of physics stay the same, just as they did when the roles of every proton and every neutron in the Universe were swapped. How does the Universe take note of the local symmetry transformation? Through, in this case, the strong force itself. The fundamental forces of nature are therefore deeply involved with the basic symmetries — not just global symmetry changes, but local ones.

There are many ways in which symmetry changes can occur, but it turns out that the symmetries which underlie the laws of physics are the simplest kind, mathematically speaking. They are called gauge symmetries, and they also have local symmetry, which, it turns out, restricts their properties and makes it possible to calculate their effects.

The term 'gauge' is simply a label mathematicians use to describe a property of the field. The term was introduced into his context shortly after World War One, by the German mathematician Hermann Weyl, who was trying to develop a unified theory combining electromagnetism (Maxwell's equations) and gravity (General Relativity). A gauge transformation is one which changes ('regauges') the value of some physical quantity everywhere at once, and the field has gauge

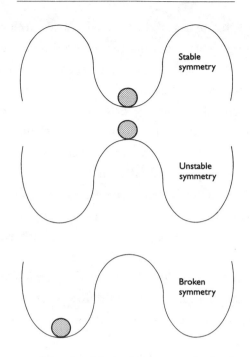

Figure 2.7   The breaking of symmetry in fundamental processes can be understood in terms of a ball in a valley. With one valley, the ball is in a stable, symmetric state. If there are two valleys, even though they are symmetric in themselves, if the ball is present the symmetric state is unstable and the tiniest nudge will send the ball rolling one way or the other, breaking the symmetry.

symmetry if it stays unchanged after such transformation. A good example is provided by the imaginary system of electric charges I have already referred to. If we set such a system up in a real laboratory, and measured all the forces between the charges, we would find that it made no difference to these forces if we charged up the whole laboratory to a high voltage.[20] The only thing that matters is the difference between the charges — and that is why a mouse can run quite happily along the live rail in the subway. All of the mouse is at the same voltage, and no currents flow. The problems arise for a person who touches the rail with one hand and the ground with some other portion of the body, allowing electric current to flow across the potential difference.

So the electric forces between particles are invariant if the potential (the voltage) on every charge is increased by the same amount at the same time. This gauge invariance is another kind of symmetry, one that is shared by the gravitational field. But what happens if only part of the system of charges is raised to a higher electric potential? Now electric currents begin to flow, just as when someone, or something, falls on a live rail in the subway. The moving electric charges create a new field, a magnetic field, which can be described in terms of a magnetic potential, analogous to an electric potential. And the magnetic field restores the symmetry of the equations describing the system. If we imagine making any kind of complicated change in the electric potential anywhere in the lab, or in the Universe, raising it here and lowering it there, we can always cancel out the effects of these changes by making compensatory changes in the magnetic potential, lowering it there and raising it here. Electromagnetism, the theory that

---

[20] And this isn't just an imaginary 'thought experiment'; it can be, and has been, done.

includes both electricity and magnetism, is therefore invariant under local gauge transformations. Indeed, Maxwell's equations describe the simplest kind of field which obeys both this symmetry invariance and the equations of special relativity.

This kind of symmetry is very deeply connected with the equivalence principle in general relativity. Einstein taught us that accelerations could always be cancelled out by gravity. Accelerations represent force. As Newton taught us, force is equal to mass times acceleration. In a laboratory moving at constant velocity through space, there is no change in the gravitational potential from one end of the lab to the other, and we have a situation like our electrical set up with a uniform baseline voltage. Experiments in such a laboratory obey Newton's laws perfectly. They show a symmetry, similar to the symmetry shown by the system of charges in the earlier example. On Earth, there is a gravitational potential difference from the top of the lab to the bottom, because of the Earth's gravity. That is equivalent to our electrical set-up with one end of the lab charged to a higher voltage than the other. The symmetry is no longer there.

If you keep the imaginary lab in space, but jiggle it about by firing rocket motors from time to time, the effects show up in the laboratory as mysterious forces affecting the trajectories of particles. Those forces are exactly equivalent to the ones produced by gravity. To make the lab fly in a circle, for example, far from any large mass, you would have to apply a constant push, and an occupant of the lab could deduce that it was moving in a circle by measuring the forces inside. But if our spacelab is in orbit around the Earth, the forces that 'ought' to show up because it is moving in a circle, not in a straight line, are precisely cancelled out by the force of gravity from the planet below. It is in free fall. In principle, just as you could make the magnetic potential compensate for changes in the electric potential, so you could make a changing gravitational potential to cancel out even the most violent buffeting produced by the rocket motors. Putting it another way, you could, in principle, arrange lumps of matter (planets, stars, black holes or whatever) around the spacecraft so that it followed the most bizarre wiggly line trajectory through space, but was always in free fall, just as a spaceship orbiting the Earth is in free fall in a circular trajectory. It doesn't matter that this is not a practical proposition; the point is that the symmetry is built into the equations. The gravitational field is invariant under local gauge transformations.

But all we can ever 'know' about the forces of nature is, indeed, the way in which they affect motion, deflecting an electron from its trajectory here, nudging a proton there, and so on. The other forces of nature play out the same role at the particle level that gravity does in the Universe at large, providing a means to cancel out disturbances caused by local symmetry transformations. In the quantum physics description of electromagnetism, QED, the force is equivalent to an exchange of photons between charged particles. And the changes in the particles and their associated fields cancel out, to ensure local gauge symmetry, if and only if the photon is a particle with one unit of spin and zero mass. The existence of the photon, with just these properties, is seen by physicists today as a requirement of gauge symmetry, or, depending on your point of view, as confirmation that the gauge symmetry approach is the key which will unlock the secrets of nature. So what happens when that approach is taken over from QED into the description of the weak and the strong fields?

Chapter Three

In Search of
Superforce

The idea of finding one mathematical description which would include all of the forces of nature has been the Holy Grail of physics from the moment Einstein came up with a field theory of gravity, the General Theory of Relativity. Early attempts started out by trying to unite gravity and electromagnetism into one theory — in the 1920s, after all, those were the only two fully worked out theories that physicists had to play with. Those attempts failed, although some of the techniques developed in the construction of those failed attempts at unified field theories have been revived, and are proving remarkably successful, in the context of physics in the second half of the 1990s. Gravity is the weakest of the four forces of nature, and the hardest to reconcile with the other three. Although it has a very long range (the particle of the gravitational field, the graviton, is, like the photon, massless), and therefore dominates the Universe at large, gravity is easily overpowered by any of the other forces when they are working over a range where each can be effective. It takes the mass of the whole Earth to hold a scrap of paper on my desk down, with a weight of less than a gram. But I can lift that piece of paper up, against the gravitational pull of the whole Earth, simply by rubbing a plastic pen on my woollen sweater to build up an electric charge on the pen, and holding the charged pen over the paper. The electric force which the pen exerts on the paper then makes it jump off the desk. Gravity really is a *very* weak force indeed. The only reason electric forces don't dominate the Universe is that almost everywhere the positive and negative charges are in balance, so that there is no net charge left over to influence distant stars and galaxies. The weak and strong nuclear forces are also much stronger than gravity, but fortunately have only limited ranges, because they are mediated by particles (field quanta) with mass. In round terms, the strong force is 1000 times stronger than the electric force, and 100,000 times stronger than the weak force (so the electric force is about 100 times stronger than the weak force). But the strong force is 10^{38} times stronger than gravity, and it is therefore no surprise to find that developing a unified description of the strong, weak and electric forces is far easier than trying to find a unified theory pairing up gravity with any one of the other three.

When physicists appreciated that they had four fundamental forces to deal with, not two, the problem of unified field theory looked more daunting than it had in the 1920s. A few researchers, Einstein prominent among them, kept plugging away at developing sets of equations that might describe a unified theory which included gravity, electromagnetism and the others in one package, but even Einstein had little success, although he spent most of the last thirty years of his life trying to unify electromagnetism and gravity. When success did begin to come, it came by starting from the opposite 'end', as it were, to Einstein. He started out with the force of gravity, which dominates the Universe on the large scale. But today, adding gravity in to the unified theory is seen by physicists as the *last* piece of the puzzle to tackle. Instead, they have proceeded piece by piece, starting with the forces that dominate atoms — indeed, starting with the two forces closest to each other in strength — and working outwards into the Universe. The weak force was first given its own 'proper' field theoretical model, then added to the electromagnetic force to provide a unified electroweak model. Today, there is also a 'proper' gauge theory of the strong force, and clear indications of how to add that strong force in to the electroweak field as well, to provide a Grand Unified Theory or GUT. There is no *unique* GUT which does unify the first three forces, but the family properties of the kinds of model that almost certainly include such a unified theory have been outlined. And there is real hope of bringing gravity within the fold.

These ultimate developments involve an understanding of particle processes that go on at very high energies, energies which are equivalent to densities of matter far greater than the density of matter in an atomic nucleus. So the steps toward a unified theory of fields are also, in a real sense, steps back towards the moment of creation, the superdense, superenergetic state known as the Big Bang, in which the Universe was born. The theories tell us what conditions in the Universe were like during the first split-second of creation, as I explain in my book *In Search of the Big Bang*. This is, perhaps, as clear an indication as any that these theories, involving quarks and leptons, really are reaching some fundamental level of physics. The success of these models in explaining the early evolution of the Universe — the match between particle physics and cosmology — is one of the best pieces of evidence we have that those particle and force theories are developing along the right lines.

If the new theories stand up, as they seem to be doing so far, physics will soon have achieved its fundamental goal of describing everything by one set of equations, and that will imply an understanding of the Big Bang itself from the moment of creation to the end of time. The one unified field — what Paul Davies has called the 'superforce' — is the key to understanding not only how the world works today, but how it got to be the way we see it today. And the successful search for the superforce follows a trail that began in 1954, just one year before Einstein died, when a Chinese-born physicist working in the United States with an American colleague published a paper applying the idea of a local gauge theory to the problem of the strong force. Their model was not particularly successful as a description of the strong force, but marked a conceptual breakthrough that encouraged other researchers to tackle other problems using similar techniques. Ironically, the first fruits of this attack on the *strong* force turned out to be a better understanding of the *weak* interaction.

Electroweak Unification

Chen Ning Yang had been born in Hefei, China, in 1922. His father was a professor of mathematics, and Yang himself studied at the Chinese universities in Kunming and Tsinghua, where he obtained an M.Sc., before moving to Chicago in 1945 to work for his Ph.D., which was awarded in 1948, under the guidance of Edward Teller. He then spent a further year at Chicago as an assistant to Enrico Fermi, and in 1949 he joined the staff at the Institute for Advanced Study, in Princeton, where he stayed until 1965. Yang was interested in the possibility of modelling a field theory of the strong interaction along the same lines as QED, and he worked intermittently on the problem, with only limited success, from the time he was in Chicago until 1954. Then, he spent a year away from Princeton, at the Brookhaven National Laboratory, where he shared an office with the theorist Robert Mills.

Mills and Yang together were able to construct a gauge-invariant field theory of the strong interaction. The symmetry that is important in the Yang–Mills theory is the isotopic spin symmetry, which I have already mentioned. In such a description of nucleons, protons and neutrons are represented by vertical and horizontal arrows respectively in a mathematical space; and if there is a local symmetry then that means that it is permissible to vary the isotopic spins of individual nucleons at different places in the Universe, and at different times. In other words, there are interactions which change individual protons into neutrons, and vice versa. The simpler global symmetry of course, 'only' allows us, in imagination, to change *all* the neutrons into protons and all the protons into neutrons, and all at the same time.

Just as with other theories of this kind, the way the symmetry is preserved when we are allowed to make local changes in the field is by adding in something else to counterbalance the changes we are making. In the Yang–Mills theory, the laws of physics can only be made to stay the same even when arbitrary changes in isotopic spin are made by including *six* vector fields. Two of these fields are equivalent, mathematically, to the ordinary electric and magnetic fields, and together they describe the photon, the carrier of the electromagnetic force. The other four fields, taken in two pairs, describe two new particles, which are similar to the photon but carry charge, one positive and one negative. And the interactions involving all these particles, as represented in the theory, were horribly complicated.

It was clear that this approach to an understanding of the strong interaction was, at the very least, incomplete. For a start, none of the 'photons' had any mass, so they would have infinite range, whereas in fact the strong force is the one with the shortest range of the four classic forces, so its carriers ought to have relatively large masses. But the ideas underlying the model were, and are, very interesting indeed. At a simple level, two oppositely charged 'photons' could be imagined binding together, like a proton and an electron, to make an 'atom' of strong field. At a rather deeper level, one of the fundamental discoveries, which had important consequences in the development of later theories of the four interactions, was that because of the presence of the charged photons the order in which a series of transformations is applied to a fundamental particle can make a crucial difference to the state it ends up in.

That sounds complicated, so let's take it step by step. The state of an electron, for example, can be changed by absorbing or emitting a photon of light. If the electron first absorbs and then emits the photon, it will end up in the same state as if it first emitted and then absorbed the photon (assuming it starts out the same in each case, and that all the photons are identical). The order in which the interactions take place doesn't matter, and QED is therefore said to be an Abelian theory.[1]

Ordinary numbers work like this. We all know that 2×4 is the same as 4×2, and that $6 + 7$ is the same as $7 + 6$. The numbers are said to commute, and in general we can write $A \times B = B \times A$. But in quantum physics, this is generally not the case. It turns out that $A \times B$ is not equal to $B \times A$, and the variables are said to be non-commutative, or non-Abelian. The same thing happens with the charged 'photons' of the Yang–Mills theory. If a hadron is changed by a local rotation of the isotopic spin arrow, and then it is changed a second time

---

[1] After Niels Henrik Abel, the Norwegian mathematician who lived from 1802 to 1829, and made major contributions to the branch of mathematics known as group theory. His early death was a great blow to nineteenth century mathematics.

by a second, different, isotopic spin rotation, the state it ends up in depends on the order in which the changes were made. The Yang–Mills theory is a non-Abelian local gauge theory, and it turns out that all of the fundamental fields are described by non-Abelian gauge theories — even electromagnetism, as we are about to see, is part of a bigger, non-Abelian theory.

All this may sound very deep and technical indeed. But you can demonstrate non-Abelian transformations simply by using the book you now hold (or another one). Place the book flat on the table in front of you, with the front cover showing and the right way up. If you rotate the book through 90 degrees by lifting the end furthest from you (the 'top' of the book), the book will be standing upright, with the cover facing you. Now look down on the top of the book and rotate it by 180 degrees. It will be left upright, the right way up, with the *back* of the book facing you. Now try again, starting from the same place as before (book flat on the table, face up) and doing the same rotations in the opposite order. First, rotate the book 180 degrees, so it is flat on the table but the title is upside down. Now rotate it 'upwards' by 90 degrees, lifting the far end. You end up with the *front* of the book facing you, but with the book upside down. It is the same book, with the same amount of energy, but it is in a different state. You have carried out a couple of non-Abelian transformations of the book.

Although theorists in the mid-1950s knew full well that a little more work was needed on the Yang–Mills theory, basic ideas like these were interesting and encouraged new lines of thought — they were certainly interesting enough to justify the publication of the paper setting out the theory, in 1954.[2] It was to take twenty years for theorists to develop this approach into a satisfactory theory of the strong force, and they made very little progress until the late 1960s, when quarks were recognized as the fundamental entities involved in the interactions, and gluons as the carriers of the true strong force. But meanwhile the ideas were taken over into a theory of the weak interaction, and then into the electroweak theory, uniting electromagnetism and the weak force.

Julian Schwinger was something of a child prodigy in mathematics. He was born in 1918, and entered the City College of New York at the age of fourteen, then transferred to Columbia University, where he gained his B.A. degree at the age of seventeen, and a Ph.D. three years later. He worked with Robert Oppenheimer (the 'father of the atom bomb') at the University of California, then at the University of Chicago and at MIT, before he joined the faculty of Harvard University in 1945. A year later, at the age of twenty-eight, he became one of the youngest full professors ever appointed at the august institution. Schwinger made major contributions to the development of QED, and in 1965 he shared the No-

---

[2] Incidentally, scooping another theorist who was thinking along similar lines independently. Ronald Shaw was a student working at the University of Cambridge under the supervision of Abdus Salam; he came up with a very similar model to Yang and Mills, but after their paper appeared in the *Physical Review* in October 1954 (volume 96 page 191), he didn't bother to attempt to get his own version published.

bel Prize in Physics with Richard Feynman and Shin'ichiro Tomonaga, of Tokyo University, for this work.[3]

So Schwinger had the ideal background to pick up the Yang–Mills idea and apply it to the weak force and to electromagnetism. The rules of the game are slightly different with the weak interaction. In beta decay, for example, a neutron is converted into a proton, so the isotopic spin (isospin) symmetry is disturbed. But at the same time, in such an interaction, a neutrino is converted into an electron (or an antineutrino and an electron are created together, which is the same thing) so there has been a transformation in the lepton world analogous to the isospin change in the hadron world. This leads to the idea of 'weak isospin', a quantum parameter like isospin, but one which applies to leptons as well as to hadrons. In 1957, Schwinger took over the non-Abelian local gauge theory developed by Yang and Mills for the strong force, and applied it to the weak force and electromagnetism (QED) together. Like the Yang–Mills theory, his version had "new" vector bosons, one without charge and the other two carrying charge. And, like Yang and Mills, he identified the uncharged field quanta with photons. But, unlike Yang and

---

[3] Tomonaga, who was born in 1906 and died in 1979, worked in isolation from the American scientists and published his results first, in 1943. Feynman and Schwinger made their independent contributions to QED just after the war, and Tomonaga's work became known to the English speaking world in 1947. The three of them had arrived at the same model by three different routes, which was in itself a strong indication that the model they came up with described some fundamental feature of nature.

Mills, in Schwinger's treatment the two charged vector bosons were regarded as the W^+ and W^-, the carriers of the weak force. There was still the problem with masses. Masses had to be added in to the theory for the W particles more or less by hand, on an *ad hoc* basis. But this theory, in spite of its obvious flaws, again raised interesting new ideas. It implied that the weak force and the electromagnetic force were 'really' the same strength as each other, in some sense symmetric, but that this symmetry had got lost, or was broken, because the W particles had mass (and therefore a limited range) while the photon had none (and therefore has infinite range).

This led to two lines of development in field theory. Sidney Bludman, of the University of California's Berkeley campus, took up the links with Yang–Mills theory, and pointed out in 1958 that the weak force *alone* could be described by a local, non-Abelian gauge theory with three particles, the W^+, the W^-, and a third vector boson, with zero charge, called the Z^0, or just Z. This left electromagnetism out of the picture for the time being, but carried with it the implication that there ought to be weak interactions that involved no change in electric charge — ones that are mediated by the Z particle, and are known as neutral current interactions. All these field quanta were still massless in Bludman's model, so the model was far from being realistic. But perhaps it was less far from the 'answer' than earlier models had been.

Meanwhile, Sheldon Lee Glashow, a physicist who had been born in the Bronx in 1932 and graduated from Cornell University in 1954, had been studying for his Ph.D. at Harvard under the supervision of Schwinger. He found a way to take Bludman's variation on the theme and to combine it with a description of electromagnetism, producing a model, which he published in 1961, that included both a triplet of vector bosons to carry the weak field *and* a single vector boson to carry the electromagnetic force. The only immediate benefit of this approach was that it proved possible to ensure that the way the singlet and triplet mixed together produced one very massive neutral particle, the Z, and took all the mass away from the other one, the photon, instead of having two neutral particles which each had mass. But you still had to put the masses in by hand, to destroy the symmetry between electromagnetic and weak forces in the basic equations, and, worst of all, the theory did not seem to be renormalizable, but was plagued by the kind of infinities that crop up in QED but are there removed by suitable mathematical sleight of hand. The mathematical sleight of hand needed to put mass in to the early electroweak models made it impossible to carry out the renormalization trick as well.

At the same time, starting out in the late 1950s and continuing into the early 1960s, the Pakistani physicist Abdus Salam and his colleague John Ward were developing an electroweak theory very similar to the one proposed by Glashow. Salam was born in Jhang, in what is now Pakistan, in 1926. After attending Punjab University he went on to Cambridge, where he was awarded his Ph.D. in 1952, and taught in Lahore and at Punjab University until 1954, when he returned to Cambridge and, among other things, supervised the work of student Ronald Shaw. The subjects chosen by students for investigation usually reflect the interests of their supervisors, and Shaw's work was no exception. Salam was indeed interested in gauge theories of the basic forces of nature, along the lines of the Yang–Mills theory. In 1957 he took up a post as Professor of theoretical physics at Imperial College in London, and in 1964 he was the moving force behind the establishment of the International Centre for Theoretical Physics in Trieste, an institute which provides research opportunities for physicists from the developing countries. Until his death in 1997, Salam was Director of the Centre in Trieste, and spent some of his time there and some at Imperial College.

The Salam–Ward variation on the electroweak theme (Ward, a British physicist, worked at several US institutions in the 1960s, including Johns Hopkins University) suffered from the same defects as Glashow's version — the masses had to be put in by hand, and, largely as a result of this, it was impossible to renormalize the theory. The first step towards solving this problem was taken in 1967, when Salam and, independently, American physicist Steven Weinberg, found a way to make the masses of the weak vector bosons appear naturally (well, almost naturally) out of the equations. The trick involved spontaneous symmetry breaking, and once again it depended upon ideas that had been developed initially in the context of the strong field.

Actually, you can understand symmetry breaking quite easily in the context of the weakest field, gravity. To an astronaut in free fall in a spacelab, there is no special direction in space. If the astronaut lets go of a pen, it floats off in any direction the astronaut pushes it. All directions are equivalent;

there is a basic symmetry. On the surface of the Earth, things are different. If you give a pen a slight push in any direction and let go of it, it always falls the same way, downwards. 'Downwards' means towards the centre of the Earth. Drop a pen at the North Pole and it falls downwards; drop a pen at the South Pole and it falls downwards. But the two 'downwards' are opposite to one another. The basic symmetry is hidden, or broken by the Earth's gravitational field.

Another form of hidden symmetry applies to a common bar magnet, which always prefers to line itself up pointing north–south, even though the basic equations of electromagnetism are symmetric. This form of hidden symmetry was discussed half a century ago, by the physicist Werner Heisenberg, the same Heisenberg who derived the uncertainty relations for the first time. But the easiest example to understand involves gravity once again. Imagine a perfectly smooth, perfectly symmetrical surface shaped like a Mexican hat, with an upturned brim. If the 'hat' is resting level on a horizontal surface then it is completely symmetrical in the Earth's gravitational field. Now imagine placing a small, round ball on top of the hump in the middle of the hat. Everything is still perfectly symmetrical, as long as the ball doesn't move. But we all know what will actually happen in such a situation. The ball is unstable, balanced at the highest point of the hump, and will soon roll off and fall down one side of the hump to rest in the rim of the hat. Once this happens, the hat and ball together are no longer symmetric. There is a special direction associated with the system, a direction defined by a line pointing outwards from the centre of the hat through the place where the ball rests on the rim. The system is now stable, in the lowest energy state that it can easily reach, but it is no longer symmetric. It turns out that the masses associated with the field quanta in a Yang–Mills type of theory can arise from a similar symmetry breaking involving the abstract 'internal space' in which the arrows of isospin point.

The idea gradually brewed up in the 1950s and 1960s from the work of several mathematical physicists, but it came into full flower with the work of Peter Higgs, at the University of Edinburgh, between 1964 and 1966. Higgs had studied at King's College, in London, from 1947 onwards, and received his Ph.D. in 1954. He took up a post at Edinburgh in 1960. Although the line of thought behind the mechanism he proposed is too complex to detail here, its implications can be understood in terms of the same language that we are now becoming familiar with. Higgs proposed that there must be an

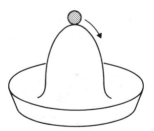

Figure 3.1 'Mexican hat' symmetry. The ball on top of the hat represents an unstable symmetry (see *Figure 2.7*) that can be broken by the ball rolling off in any direction.

extra field added in to the Yang–Mills model, one which has the unusual property that it does not have the least possible energy when the value of the field is zero, but when the field has a value bigger than zero. The electromagnetic field, and most other fields, have zero energy when the value of the field is zero, and the state in which all fields have minimum energy is what we call the vacuum. If all fields were like the electromagnetic field, that would be the same as saying that in the vacuum state all fields are zero. But the Higgs field has a non-zero value even in its state of minimum energy, and this gives the vacuum itself a character which it would not otherwise possess. Reducing the Higgs field to zero would actually involve putting energy into the system.

The implications of this are profound. In terms of isospin, the Higgs field provides a frame of reference, a direction against which the arrow that defines proton or neutron can be measured. A proton can be distinguished from a neutron by comparing the direction of its isospin arrow with the direction defined by the Higgs field. But when the isospin arrow rotates during a gauge transformation, the Higgs arrow rotates as well, so that the angle stays the same. The angle that used to correspond to a proton now corresponds to a neutron, and vice versa. Without the Higgs mechanism, there would be no way to tell the difference between protons and neutrons at all, because there would be nothing to measure their different isospins against. All that can be measured is the relative angle between the isospin and the Higgs arrow, not any absolute orientation of isospin. And the Higgs field does this even though the field itself is a scalar, which has only magnitude and does not point in any preferred direction at each point of 'real' space.

The effect of all this on the vector bosons is dramatic. There are four scalar Higgs bosons required by the field theory, and as we already know the basic Yang–Mills approach gives three massless vector bosons. When the two elements are put together, three of the Higgs bosons and the three vector bosons merge with one another — in the graphic terminology used by Abdus Salam, the vector bosons each 'eat' one of the Higgs particles. And when this happens the vector bosons gain both mass and a spin corresponding to the spin carried by the Higgs bosons. Instead of having three massless vector bosons and four Higgs particles, the theory predicts that there should be three observable vector bosons which each have a definite mass, plus one scalar Higgs boson, which also has a large mass but whose precise mass cannot be predicted by the theory. The Higgs field breaks the underlying symmetry in just the right way to fit in with what we observe. At the cost of one extra undetected particle, mass appears naturally in all the variations on the Yang–Mills approach.

Higgs himself had been working in the context of the strong field. But his ideas were soon taken over into the developing electroweak theory. First off the mark was Steven Weinberg, in 1967. Weinberg had been an exact contemporary of Glashow (although six months younger, having been born in May 1933), both at the Bronx High School, from which he graduated in 1950, and Cornell University, where he graduated in 1954. But then he followed a different path to end up with a model very similar to Glashow's description of the electroweak interaction, but with the bonus of a Higgs-type mechanism included. By 1960, he had arrived at Berkeley, where he stayed until 1969 before moving on first to MIT and then, in 1973, to Harvard. Weinberg's approach to electroweak unification was largely his own, but drawing upon the same culture — the same background pool of knowledge in physics — that Glashow and Salam were drawing on. His interest in the weak interaction went back to his Ph.D. work in Princeton, and in the 1960s he worked towards an equivalent of the Higgs mechanism in his own way. His electroweak model, including masses for the vector bosons generated by spontaneous symmetry breaking, was submitted for publication in October 1967, and appeared in the journal *Physical Review Letters* before the end of the year.[4]

Salam heard about the Higgs mechanism, from a colleague at Imperial College, a few months before Weinberg submitted that paper for publication. He took the electroweak model he had developed with Ward and added the Higgs mechanism to it, giving essentially the same basic model that Weinberg developed, with masses now occurring naturally, and he gave a series of lectures on the new model at Imperial College in 1967, followed by a talk at the Nobel Symposium in May 1968, later published in the symposium proceedings.

In due course, Glashow, Salam and Weinberg jointly received the Nobel Prize in Physics for their roles in creating a

---

[4] Volume 19, page 1264.

unified electroweak theory, a step as important as Maxwell's development of a unified electromagnetic theory a century before.[5] But 'in due course' was not until 1979. It took some time for even most theorists to appreciate fully the significance of the Weinberg–Salam model, because it wasn't until 1971 that a Dutch physicist, Gerard 't Hooft, showed that this version of the electroweak theory was, indeed, renormalizable. And then, in 1973, experiments at CERN came up with evidence of the elusive neutral current interactions that the theory predicted, interactions mediated by the neutral Z particle. It was the normalization of gauge theory by 't Hooft that led to the explosive development of field theory in the 1970s, to a theory of the strong interaction, and even to an understanding of the earliest moments in the life of the Universe itself.

Gauge Theory Comes of Age

The way I have told the story of the development of gauge theories in the 1950s and 1960s may seem logical and orderly, the onward, inexorable march of science. But that is only true up to a point. The path followed by Weinberg, Salam and the others in the 1960s, in particular, was then very much a byway of science. The theorists who dabbled in such things as non-Abelian local gauge theories were as much mathematicians as

---

[5] Don't shed too many tears for Yang, left out of the 1979 award. He already had a share in the Physics Prize, awarded in 1957, for another key contribution to particle physics theory, which comes into the story of the early Universe which I have described elsewhere.

they were physicists, and they were often interested in the equations and the symmetries as much for their own sake as for their bearing, if any, on the real world. It is only with hindsight that we look back from the late 1990s and see this single thread of the whole tapestry of science as being particularly important, and leading on to greater things. And that is shown very clearly by the way in which Weinberg's paper on electroweak unification, published in 1967, was almost totally ignored for four years.

The fate of scientific papers published in the major journals, which Weinberg's was,[6] is monitored in the pages of a publication called the *Scientific Citation Index*, which lists, each year, the number of times that each paper is referred to by the authors of other papers published in the major journals. In 1967, 1968 and 1969 *nobody* (not even Weinberg himself) referred to this paper in print. In 1970, there was just one citation; in 1971 there were four; in 1972, sixty-four; and in 1973 there were 164.[7] The sudden upsurge following 1971 was entirely due to the breakthrough achieved by Gerard 't Hooft, when he showed that gauge theories in general, and the electroweak theory in particular, were renormalizable.

Progress towards such an achievement had been slow and

---

[6] Salam's 1968 paper was not even published in a major journal of physics, but in the more obscure pages of the Nobel Symposium proceedings, where it didn't even fall within the net of the citation index. So we have no comparable figures for that paper, but since its publication was so obscure it can hardly have received wider attention than Weinberg's paper in *Physical Review Letters*.

[7] Figures reported by Pickering, page 172.

painful, and there is no point here in recounting all the detours into blind alleys on the way. So, once again, the story may seem straightforward and uncomplicated; but remember, once again, that this is only because we have the benefit of hindsight.

This thread of the story begins with the work of another Dutch physicist, Martin Veltman, who was born in 1931 and studied at the University of Utrecht and spent five years at CERN before taking up the post of Professor of Physics back at his old university. Veltman developed for himself, by a roundabout route, a set of gauge equations equivalent to the Yang–Mills model of fields, and although confused by a discussion with Richard Feynman in 1966, in which Feynman advocated a different approach to the problems of particle physics, he eventually decided to follow up a suggestion made by John Bell, a British physicist working at CERN, that the best way ahead would be through the development of a Yang–Mills model for the weak interaction. He tackled the problem in his own way, using the path integral approach pioneered by Feynman but which few physicists then took very seriously as a practical tool.

The obvious major problem with all models of the Yang–Mills kind was the way in which infinities appeared, and could not be cancelled out. In the middle of the 1960s, it looked as if there might be no way of removing these infinities — that the theories were *in principle* not renormalizable. But with the aid of the electronic computers that were becoming increasingly important to such work as the 1960s drew on, Veltman was able to find ways in which many of the infinities could be cancelled, and to show that it might after all be possible, in principle, to renormalize the theory fully. He spent years laying the groundwork, backing and filling and covering an enormous amount of ground, but never quite achieved his goal of the renormalization itself. That was left to the next person to pick up the torch.

Gerard 't Hooft was born in the Netherlands in 1946. He joined the University of Utrecht as an undergraduate in 1964, and began full time research, for his Ph.D., under the supervision of Veltman in 1969. The problems he chose to tackle, and the way in which he chose to tackle them, were both far from the mainstream of science. For a start, he was interested in gauge theories, which were completely out of fashion. And then, following Veltman's lead, he decided to tackle the gauge theory problems using Feynman's path integral approach. Taking over many of Veltman's techniques, 't Hooft was able to show, in a paper published in 1971, that *massless* gauge theories are indeed renormalizable. This was an excellent achievement for a student just starting out in research, but the really important problem, of course, was to renormalize the theories which included massive particles, the Ws and the Z, the intermediate vector bosons that were thought to carry the weak force. Much later, Veltman told Pickering of a conversation with 't Hooft in early 1971, a conversation so striking that it was burned in his memory and could be repeated more or less verbatim more than ten years later. Translated into English, it went something like this:

Veltman: I do not care what and how, but what we must
have is at least one renormalizable theory with massive
charged vector bosons, and whether that looks like nature
is of no concern, the details can be fixed later.
't Hooft: I can do that.
Veltman: What do you say?
't Hooft: I can do that.
Veltman: Write it down and we will see.[8]

't Hooft did write it down, and Veltman saw that he had indeed cracked the problem. The resulting paper was published
in the journal *Nuclear Physics* before the end of 1971 (volume
B35, page 167), and 't Hooft was awarded his Ph.D. in March
1972. By then, the transformation of particle physics and the
restoration of gauge theories to centre stage was already occurring, thanks to the way the word of this work by an obscure
student tackling an obscure problem with an obscure technique was spread in the United States by physicist Benjamin
Lee, who had been a visitor at Utrecht in the summer of 1971,
and returned to the US armed with copies of both of 't Hooft's
1971 papers. Lee both confirmed the validity of 't Hooft's
work and translated it into more conventional mathematical
language in his own paper, published in 1972. It was Lee's paper that persuaded theorists such as Weinberg to take the
work seriously, and convinced them that gauge theories of
the electroweak interaction involving symmetry breaking and
the introduction of mass through the Higgs mechanism were

---

[8] Slightly paraphrased from Pickering, page 178.

indeed renormalizable. Gauge theory had come of age, at least as far as the theorists were concerned.

In 1973, experiments at CERN which involved shooting beams of high energy neutrinos through a huge bubble chamber called Gargamelle produced evidence of interactions involving the elusive Z particle. The tracks in the bubble chamber showed that an antineutrino or a neutrino could interact with an electron just as the electroweak theory predicted, with the Z^0 mediating the interaction (because no charge is transferred, this is called a 'neutral current' interaction). Further experiments confirmed the plausibility of this interpretation of events — after analysing three million photographs of events occurring inside Gargamelle, the physicists found 166 examples of interactions best explained in terms of neutral currents. Now, the experimenters too were persuaded that the electroweak gauge theory was the best theory of interactions involving leptons and photons.

The significance of the finds, coupled with 't Hooft's renormalization of the electroweak theory, was so profound that Weinberg, Salam and Glashow were awarded the Nobel Prize in 1979, even though at that time there was still no direct evidence that the Ws and the Z existed. But the theory had predicted not only that these particles must exist, but also the masses they should have. The Ws ought each to weigh in at about 92 GeV (a little less than 100 times the mass of a proton), and the Z^0 should have a mass of about 82 GeV. In order to make such particles and watch them decay, you need a particle accelerator which can put at least this much energy into collisions. Just such an accelerator was built at Geneva, by

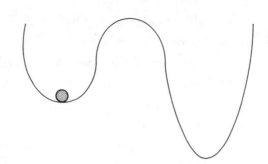

Figure 3.2 Nature is not always blessed with underlying symmetry. In this case, the situation is asymmetric even without the ball being present (cf. *Figure 2.7*). Now, the ball can be in a state which represents a *local* minimum in energy, but is not the most stable *possible* state — rather like the alpha particle in an unstable nucleus (see *Figure 1.4*). At high energies, far above the two valleys, the asymmetry is not apparent; at low energies, there is a choice of states. Conceivably, as in the example shown here, the ball will eventually tunnel into the lowest energy state.

CERN, to smash a beam of protons head on into a beam of antiprotons. And by the early months of 1983 the accelerator produced clear evidence for their existence of W and Z particles, with masses very close to the predicted masses, which are produced in the collisions and then decay into energetic electrons and other particles.[9] No doubt the discovery, confirming the basis of their 1979 award, came as a quiet relief to the

---

[9] Details of the search for these particles can be found in Christine Sutton's book *The Particle Connection*, which describes how such energetic collisions are achieved (a saga in itself) as well as what the observations imply.

Nobel Committee. Quick off the mark, they gave the 1984 Physics prize to Carlo Rubbia, the head of the CERN team involved in the work.

The significance of these masses for the unification of the forces in the Big Bang is easy to see. When the energy density (temperature) of the Universe was great enough, particles with masses of just under 100 GeV could appear spontaneously, in particle-antiparticle pairs. And instead of a carrier of the weak interaction being able to come into existence only for the brief instant of time allowed by the uncertainty principle, the supply of free energy around it could make any of these virtual particles real, and give it an extended lifetime. As long as the mass of the particle was less than the energy available, it could live forever, like the photon, and the distinction between photons and the Ws and Zs would be dissolved away. At high enough energies, during the early phases of the Big Bang, there is no distinction between the electromagnetic and the weak forces. The distinction only appears because we live in a cold Universe, where the symmetry is broken. The Ws and Zs began to freeze out of our Universe when its temperature fell to 10^{15} K, about one thousand-millionth of a second after the moment of creation. And that is when the electromagnetic and the weak forces started to go their separate ways — until mankind intervened, on a modest scale, recreating for a tiny fraction of a second, in a tiny volume of space inside a machine near Geneva, the conditions that had existed everywhere in the Universe one thousand-millionth of a second after the moment of creation.

By 1985, the proton-antiproton collider at CERN was

producing energies of 900 GeV, a new world record, and by the end of the 1980s these intermediate vector bosons were being produced as a matter of routine. But there is no prospect of achieving the same kind of success by creating the particles required by higher order unified theories of the forces of nature. The new theories based on the triumphs of the electroweak gauge theory tell us that the masses of these particles are far beyond the range of any conceivable man-made accelerator. The only place such energies have been available was in the early stages of the Big Bang itself. So the Universe has become the testing ground of the latest ideas in particle physics. The ideas themselves, though, owe a great deal to the theories of electromagnetism and the weak force that have preceded them.

Quarks With Colour

In the middle of the 1960s, there were two families of leptons known, each made up of an electron-like particle and a neutrino. The pairs are the electron and its neutrino, and the muon and its neutrino. When the idea of quarks was introduced, only three were needed to explain all of the known particles. The up and down quarks formed a pair, but the strange quark was out on its own. In fact, in the paper in which he put forward the idea of quarks, Gell-Mann did speculate that there might be a fourth quark, to pair up with the strange quark and make two quark pairs to match the two lepton pairs. But the idea was soon dropped, because there was no evidence for the existence of particles incorporating the hypothetical quark. The problem of how three seemingly identical quarks could co-exist in the same state to form a particle such as the omega minus was far more pressing and absorbed a lot of effort, from those physicists who bothered to put much effort into quark theory in the middle and late 1960s, before it was finally cracked.

Walter Greenberg, a theorist working at the University of Maryland, was delighted by the idea of quarks when it was introduced in 1964, because it provided him with a practical application for some exotic field theory ideas he had been developing for several years. Greenberg was originally interested simply in developing mathematical versions of field theory, with little or no thought of practical applications. But one of his abstract ideas, called 'parastatistics', turned out to be relevant to the quark problem. Greenberg quickly applied his abstract ideas to the new model of hadrons, and came up with intriguing results. Although his approach was very technical, it boiled down to suggesting that there might be different varieties of 'paraquarks', obeying the rules of parastatistics, and that the three seemingly identical quarks in the omega minus and some other hadrons were actually distinguished from one another by a previously unsuspected property that came in three different varieties. The idea was taken up by two theorists, Yoichiro Nambu, at the University of Chicago, and M. Y. Han, at Syracuse University. They collaborated in developing, in 1965, a version of Greenberg's approach which was rooted more obviously in the world of the experimenters, and was therefore more accessible to more physicists, than the elegant mathematics of parastatistics.

The idea underlying all of this work was that each of the known quarks could come in three varieties, which are now known as colours. The terminology is no more than a convenient labelling device, like the names 'up' and 'down'. But it enables us to understand that there is a difference between a red up quark and a blue up quark, just as there is between a red up quark and a red down quark. The mathematical equations tell us how three kinds of quark ought to interact, and they do so with elegant precision. But the heart of what they tell us can be grasped in simple colour terms, in the light of what the equations tell us. The omega minus, for example, can be thought of as made up of three strange quarks, each with the same spin, but one 'red', one 'blue' and one 'green', so that they are distinguishable and therefore not identical particles in identical states. The colours are just mnemonics, more mental crutches to help us understand. But the mathematical physicists assure us that the images conjured up by the analogy are not too misleading.

At least, they do today. This was regarded as little more than a trick, with no profound meaning, in 1965. Nambu and Han muddied the waters somewhat by elaborating their model to include more triplets of quarks in an effort to remove the need for fractional charges, but since few people were taking the ideas of quarks very seriously at the time none of this work caused much of a stir. But the idea did offer new guidelines for the behaviour of quarks, including a resolution of the puzzle of why they came only in triplets (as baryons) or in pairs (as mesons). Just by specifying a single rule that the

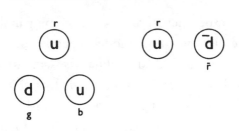

Figure 3.3 Three quarks can make up a baryon (in this case, a proton), provided they all have different colours. A quark-antiquark pair make up a pion, but must have the colour on one quark cancelled by the equivalent anti-colour on the other quark. 'Anti' is denoted by a bar over the appropriate symbol.

only 'allowed' combinations of quarks must be colourless, Nambu was able to explain the division of hadrons into these two families. Each meson, he said, must be composed of a quark of some particular colour and an antiquark of any variety but carrying the equivalent anti-colour. A red up, for example, might pair up with an anti-red up, or an anti-red down, or an anti-red strange; in each case the colour and the anti-colour 'cancelled out' in a mathematical sense. The other way to achieve a neutral state, he argued, was by mixing each of the three colours in one particle — one red quark, one green quark and one blue quark, each of them being any of the flavours up, down, or strange. Three antiquarks of different colours would achieve the same objective. But single quarks, or groups of four, for example, would carry a net colour, which seems to be forbidden.

By 1970, experimental results were coming in that seemed to be in line with this colour model of quarks, and the concept began to gain ground. And at about the same time Glashow and two of his colleagues at Harvard, John Iliopoulos and Luciano Maiani, revived the idea of a fourth quark, which Glashow gave the name 'charm', in order to tidy up the theoretical interpretation of some other puzzling experimental observations. In 1971, Murray Gell-Mann and Harald Fritzsch, who was born in Zwickau in 1943, and is now Research Professor of Physics at the Max-Planck Institute for Physics in Munich, took up the idea of colour and began to develop a field theory approach that would describe the behaviour of interactions involving particles that came in three varieties. As early as the autumn of 1972, Gell-Mann and Fritzsch were proposing that the best description of the structure of hadrons was in terms of a Yang–Mills type of gauge theory in which the triplets of coloured quarks interacted with one another through the mediation of an octet of gluons. The symmetry was more complicated, and the numbers larger, but

Figure 3.4 The colour of each quark can be thought of in terms of a knob which rotates an internal pointer with *three* settings, analogous to the isospin pointer that distinguishes protons from neutrons.

the principles were the same as those of the successful theories of QED and the electroweak force.

Once again, the basic ideas can be understood in terms of symmetry. Now, we have to imagine each baryon containing three quarks, and each quark as carrying within itself some means of selecting a colour — an internal pointer, like the up or across pointer for isospin, but now with *three* settings, corresponding to the three colours. A symmetrical global gauge transformation would be one that rotated every pointer clockwise (say) by 120°, changing the colour of every quark but leaving the laws of physics the same. A local, symmetric gauge transformation would change the pointer setting (colour) of just one quark inside one baryon, but still leave the world unchanged. And the way to restore symmetry under local transformations is, as before, to bring in new fields, corresponding in this case to the eight gluons, which are all massless (in the original version of the theory) and have one unit of spin — vector bosons analogous to the photon.

The theory became known as quantum chromodynamics, a name chosen by Gell-Mann in conscious imitation of quantum electrodynamics, and is usually referred to today as QCD. It says that any quark is free to change its colour, independently of all the other quarks, and does so by emitting a gluon, which is promptly absorbed by another quark, which suffers a colour change of exactly the kind required to cancel out the change in the first quark, and keep the hadron colourless. All hadrons are always colourless, even though the quarks within them may be undergoing kaleidoscopic changes of colour every instant. Because the gluons carry colour, their behaviour is very different from that of photons, which do not carry charge and do not interact with one another. Gluons *do* interact with each other even while they are in the process of carrying the force from one quark to another. Perhaps the strangest result of this is that although the 'strong' force is actually quite weak for a short distance (inside the proton, for example), the gluon interactions make the force stronger at *larger* distances, so that over a range of 10^{-13}cm it is strong enough to bind protons together in spite of the repulsion between their electric charges. It is like a stretching piece of elastic, fastened to a quark at each end, that only holds the quarks loosely until you try to pull them apart. Then, the more you pull, the more the elastic stretches, and the harder it pulls back, trying to keep the quarks together. In this case, the 'elastic' is a stream of gluons being exchanged between the two quarks.

If you stretch hard enough — put enough energy into a collision — then eventually the elastic will snap. But that still doesn't mean you will get a free quark emerging as a result. The energy from the interquark force goes into creating a new quark at each broken end, reminiscent of the way a bar magnet sawn in half 'creates' a new pole on each side of the break. Instead of a *single* quark emerging, there are always at least two, joined by a stream of gluon elastic — a meson. And, because the gluons carry colour, they too are forced to travel in clusters, like quarks, and cannot exist in isolation — which is why, it is suggested, no isolated gluon has been detected. Even if they are massless, gluons are unable to spread themselves about like photons. Perhaps, however, colourless bundles of gluons ('glueballs') might yet show up in experiments like those at CERN.[10]

The turning point in physics in the 1970s came in 1974, when a team at Stanford and one working at the Brookhaven National Laboratory on Long Island each discovered, almost simultaneously, evidence of a new, massive particle (now generally referred to as the psi) which was best explained as one incorporating the fourth quark, 'charm'. The discoveries led to the award of the 1976 Nobel Prize in physics to Samuel Ting, leader of the Brookhaven group, and Burton Richter, his counterpart at Stanford. The discovery was so dramatic that the announcement of the experimental results, in November 1974, is referred to by physicists today as 'the November revolution'; once *one* charmed particle had been found, the experimenters knew, in a sense, where to look for others, and soon they found a whole family of charmed particles. And that family of particles provided physicists with a testbed for QCD, which successfully predicted many details of the behaviour of the 'new' particles. With four quarks and four leptons identified, the particle world looked very neat. But there was one (last?) step to be taken.

In 1975, experiments at the Stanford Linear Accelerator suggested that there might be still another lepton, an 'electron' twice as heavy as the proton, dubbed the 'tau'; the hints were confirmed at Hamburg a year later. It is assumed, and there is very strong indirect evidence although it is not yet proven, that

---

[10] There have been claims that glueballs have been detected at CERN recently, but these claims have not yet been confirmed.

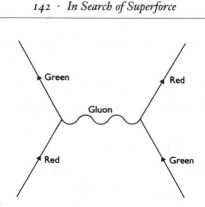

Figure 3.5 The interaction between two quarks by the exchange of a gluon can also be described using a Feynman diagram.

an equivalent tau neutrino must also exist, making six leptons, which come in three pairs. So theorists argued that there 'ought' to be two more quarks, as well, to restore the symmetry. These were called top and bottom and evidence of bottom came through in 1977; the search for the top quark came to an end in 1994, when it was identified in the debris from high-energy experiments at Fermilab, near Chicago. This really ought to be the end of the line, though. There are compelling *cosmological* reasons for thinking that there can be no more than three sets of lepton pairs in the Universe, and experiments at CERN in the late 1980s showed that the behaviour of known particles at high energies precludes the possibility of a fourth variety of neutrino, by implication ruling out a fourth variety of 'electron'. But remember that almost everything in the Universe would be the same as it is today if there were just two quarks, the up and down, and two leptons, the electron and its neutrino. The rest seems to be unnecessary triplication of effort, just one (or two) of those things that fell out of the equations and happened for no better reason than that they were not forbidden to happen.

Leptons		Quarks
e^-	=	u
ν_e	=	d
μ^-	=	c
ν_μ	=	s
τ^-	=	t
ν_τ	=	b

The combination of the electroweak theory and QCD has proved so successful in describing the particle world that it is sometimes called the 'standard model' of physics. But it is still incomplete. QCD has yet to be combined with the electroweak theory into one Grand Unified Theory, or GUT; and gravity isn't included at all. So there is plenty to keep the theorists occupied at present.[11] The main thrust of their continuing effort concerns the search for supersymmetry, and the possibility that everything in the Universe is made of string.

---

[11] There are also unresolved questions within QCD itself, for all its success so far. For instance, it may be possible, or necessary, to give the gluons mass, through the Higgs mechanism — although such a prospect is a daunting task for any theorist, faced with eight gluons to worry about and handle in one self-consistent package.

Chapter Four

Desperately Seeking SUSY

In 1991, the search for supersymmetry took a significant step forward when experiments at an accelerator called the Large Electron-Positron Collider (LEP), at CERN, filled in a missing piece of the jigsaw puzzle concerning QCD, the best theory we have of the behaviour of quarks and the strong force. The new evidence raised the confidence of theorists in QCD (uprating it from what I described in 1986, in the first edition of *In Search of the Big Bang*, as a 'good' theory to what I would now call a 'very good' theory). Although not a new step along the road to unification of all the forces of nature, but rather confirmation that an old step had indeed been made along the right path, this in turn makes theorists more confident that the right route to the superforce does involve adding QCD to the electroweak theory and to gravity.

As its name suggests, LEP collides energetic beams of

electrons head on into beams of positrons — anti-electrons. At the high energies involved in these collisions, the annihilation of an electron with its antiparticle counterpart should, the theory predicts, produce a Z^0 particle, which in turn converts initially into a quark/antiquark pair. The energetic quarks produced in these interactions should in their turn produce jets of gluons (in much the same way that accelerated electrons radiate photons), and the gluons themselves generate jets of hadrons, which (unlike the individual quarks and gluon jets) can be detected by the equipment monitoring the electron-positron collisions inside LEP. The crucial feature of QCD that has now been tested by LEP is that gluons can interact with each other at a point, something which photons cannot do. As I have explained, this is what underlies the strange way in which the strong force gets stronger as quarks try to move apart.

As a result of such interactions, it is theoretically possible, within the framework of QCD, for three jets of gluons (and therefore three jets of hadrons, at the level the CERN detectors operate at) to emerge from a single point, known as a 'three-gluon vertex'. As far as the equations are concerned, this is exactly the same as a single gluon jet splitting into two to make two gluon (and therefore hadron) jets. An example is shown in Figure 4.1. QCD predicts that about one in every hundred 'four jet' interactions like those shown in this figure will carry the imprint of a three-gluon vertex, and this is what LEP found in the spring and summer of 1991, exactly matching the predictions of the standard non-Abelian theory —

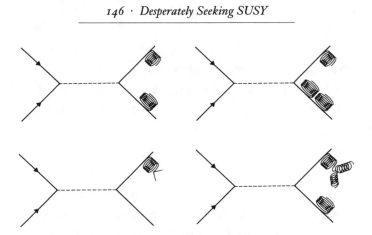

Figure 4.1 When an electron and a positron annihilate one another in a high-energy collision, they may produce a Z^0 particle (dotted line) which converts into a quark and an antiquark. These have so much energy that they radiate gluons (curly lines). The quarks and the gluons are not detected directly, but the gluons produce jets of hadrons which are detected. Evidence of QCD at work comes from 'three-gluon vertex events', like the one in the interaction shown at bottom right.

proof that the number of quark colours is indeed three and that QCD is a good description of how quarks interact. If the LEP results had failed to show the effects of the three-gluon vertex, or if the statistics of the experiments had implied that the number of quark colours was not three, the standard model of particle physics would have been in disarray. As it is, theorists can now use it with more confidence than ever before as a springboard towards ultimate unification of the forces of nature.

The Search for Supersymmetry

Even so, QCD is not yet as well established a theory as QED was even fifty years ago. It took the hindsight of the 1980s to look back and point out what turned out to be the main thread of progress in physics in the 1960s, and it will no doubt require the hindsight of the year 2020, or later, to look back on the current confusion of theoretical developments around the search for superforce and pick out the main line. I shall not attempt to advocate any one path among those now being followed as the 'best' or 'true' path; anyone who had tried that in, say, 1961 would hardly have picked out the non-Abelian local gauge theories as the best candidates to keep an eye on, let alone the idea that protons and neutrons might be made of other particles. But I can sketch in briefly the outlines of some of the most interesting *fundamental* ideas, at the level of the fundamental concepts of symmetry and gauge invariance. These ideas may underlie many different detailed theories, only one of which, we might hope, actually describes the real world. And I shall certainly give you at least a flavour of the idea that many theorists believe to be the most promising line of attack in the search for superforce as we move towards the end of the twentieth century. But whether or not the specific ideas that are front-runners today fall by the wayside, if the recent past can tell us anything, it is that powerful, simple ideas like symmetry really do help us to select good theories from bad ones.

To put things in perspective, QED is an excellent theory,

electroweak theory is very good indeed, and QCD is very good, judging by the problems that have so far been resolved and those that remain as yet unanswered. The family resemblances between the theories are perhaps the best guide that the theorists are really on the track of something more fundamental, which will unite all the forces of nature in one super theory of superforce. Electromagnetism is the simplest, and involves just one charge. The weak field is characterized by a property which has two values, isospin, and relates doublets of quarks and doublets of leptons. Quarks come in triplets, and are described by a field one step more complicated. But the same common principles underlie the singletons of QED, the doubletons of the weak field, and the triplets of QCD, and that has enabled the first two to be combined into one successful unified theory. And the colour of QCD is exactly analogous to the electric charge of QED, except that it comes in three varieties. Particles that do not carry charge cannot feel the electromagnetic field; particles that do not carry colour, the leptons, cannot feel the field of QCD.

By pushing these ideas in the same direction, many theorists have attempted to construct Grand Unified Theories that encapsulate the electroweak theory and QCD in one package. Most of these GUTs are members of the same family of theories, following a line of research pioneered by Glashow and a Harvard colleague, Howard Georgi, in the mid-1970s. Such theories each deal with particles in families of five — one such family, for example, consists of the three colours of anti-down quark, plus the electron and its neutrino. Members of these families can be changed into one another by the same kind of transformation that converts protons into neutrons and one colour of quark into another colour, equivalent to rotating a pointer which has five positions on its scale. But now we have the possibility of turning leptons into quarks, and quarks into leptons. The GUTs describe a deeper symmetry than any of the simpler theories, but at a price.

The electroweak theory needs four bosons — the photon, two Ws and the Z. The fivefold GUTs (known in mathematical shorthand as SU(5) theories) require *twenty-four* bosons. Four of these are the four already needed by the electroweak theory; eight more are the gluons required by QCD. But that still leaves twelve 'new' bosons, busy mediating new kinds of previously unsuspected interactions. Such hypothetical particles are collectively called X, for the unknown quantity, or Y. They can change quarks into leptons, or vice versa, and carry charges of $\frac{1}{3}$ or $\frac{4}{3}$. But they are very massive — so massive that their lifetimes are extremely restricted in the Universe today, and therefore they play very little part in the activity of the particle world.

According to these theories, the three forces (electromagnetism, the weak interaction, and the strong force of QCD) would have been equal to each other at energies as great as 10^{15} GeV, that is, 10^{13} (10 million, million) times the energy at which the electromagnetic and weak forces were, or are, unified. That corresponds to a time when the Universe was only 10^{-37} seconds old, at a temperature of 10^{29} K, and it means that the masses of the X particles themselves must be about 10^{15} GeV, a million, million times more than the greatest energy yet reached in a collision at the CERN proton-antiproton collider. There is no prospect of creating such conditions artificially, and that is why physicists have to look to the Big Bang for evidence that X particles ever existed. Surprisingly, though, there is a possibility of detecting a side effect of their existence here and now.

If a quark inside a proton could borrow enough energy from the uncertainty relation to create a virtual X boson and swap it with another quark, one of the quarks would become an electron (or a positron). The two quarks left over will form a meson — a pion — and the proton will have decayed. Because the X boson is so massive, its virtual lifetime is so short that it could only cross from one quark to another if they were closer than 10^{-29} cm, and this is seventeen powers of ten smaller than the size of the proton itself (10^{-17} times the size of the proton). Such very close encounters between quarks must be rare indeed. But they will happen, from time to time, and the likelihood of such events can be calculated. It turns out that for an individual proton such an event will occur once in more than 10^{30} years — probably, depending on which detailed theory you fancy, not for at least 10^{32} years. The Universe is only some 10^{10} years old, so it is no surprise to find that protons are still around and seem pretty stable. But if the chance of *one* proton decaying in *one* year is one in 10^{30}, if you have 10^{30} protons together, then there is a good chance that one of them (but you don't know which one) will decay in each year that you are watching.

Experiments have been designed to test just that — to watch large numbers of protons for months and years on end and see if any of them decay. A thousand tonnes of water contain about 10^{33} protons, and water is easy to come by. Several experiments in countries around the world have been watching for the products of proton decays in large tanks of water or lumps of iron. There is no conclusive evidence yet one way or the other, but as the years tick by the limits being set on the lifetime of the proton are up to around 10^{31} years, where things start getting interesting as far as the theorists are concerned. Perhaps some definite news, one way or the other, will be published soon.

But all is not well with the GUTs. A line of research that started out with the simple idea of symmetry in gauge theory has become ugly and complicated, with a proliferation of bosons and with the problem of what renormalization really implies still swept under the carpet, forming a bigger and harder-to-hide lump with every new force that is incorporated into the models. More quarks and leptons can be happily accommodated every time you want one, which indicates a certain lack of restraint on the part of the theories. But, embarrassingly, all of the GUTs predict the existence of magnetic monopoles, none of which have yet been found in the world we inhabit. And, indeed, since there are an infinite number of possible gauge theories it is a mystery why these particular ones should be the ones that tell us anything about the real world at all. So what might happen if we cut loose from this step by step approach, building a house of cards with one layer on top of another, and get back to the roots?

That is what Julian Weiss, of the University of Karlsruhe,

and Bruno Zumino, of the Berkeley campus of the University of California, did in 1974. GUTs surprise us by relating leptons to quarks, but they still leave bosons out on a limb as something different from material particles, merely the carriers of the forces. Weiss and Zumino said, in effect, if symmetry is a good idea, why not go the whole hog with supersymmetry, and relate the fermions to the bosons?

Stop and think about that for a minute. The distinction between fermions and bosons is *the* big one in quantum physics. Bosons do not obey the Pauli exclusion principle, but fermions do. The two seem far more unlike each other than the proverbial chalk and cheese. Can matter and force really be two faces of the same thing? Supersymmetry says yes; that every variety of fermion (every *variety*, not every individual particle) in the Universe should have a bosonic partner, and every kind of boson should have its own fermionic counterpart. What we see in our experiments, and feel the effects of in everyday life, is only half of the Universe. Every type of quark, a fermion, ought to have a partner, a type of boson called a squark; the photon, a boson, ought to have a partner, a fermion called a photino; and so on. In the same vein, there ought to be winos, zinos, gluinos and sleptons. But there is no problem in explaining where the partners have gone; at this early stage of the game, the theorists can wave their mathematical magic wands and invoke some form of (unspecified) symmetry breaking that gave the unseen partners large masses and left them out in the cold when the Universe cooled.

Claiming that there is a symmetry between bosons and fermions sounds outrageous to anyone brought up to believe in the distinction between particles and forces. But is it so outrageous? Haven't we come across something like it before? Quantum physics, after all, tells us that particles are waves and waves are particles. To a nineteenth century physicist such as Maxwell, electrons were particles and light was a wave; in the 1920s physicists learned that electrons are both particle and wave, while photons are both wave and particle. And these are the archetypal members of the fermion and boson families. Is supersymmetry really doing anything more outrageous to our commonsense view of things than taking wave-particle duality to its logical limit, and saying that a particle-wave is the same as a wave-particle? It is only because we have got away from the roots of quantum physics in the past two chapters, and, for convenience, described events in the subatomic world in terms of collisions and interactions between tiny hard particles, that supersymmetry strikes us as very odd at all. If only our minds were equipped to handle the same concepts in a more abstract form, in keeping with the quantum equations, so that we could properly understand the nature of quantum reality, where nothing is real unless it is observed, and there is no way of telling what 'particles' are doing except at the moments when they interact with one another, then supersymmetry would seem much more natural. The flaw lies in our imagination, rather than in the theory. But even with our limited imaginations we can appreciate one feature of the new theory that makes it stand head and shoulders above most candidates for the title of 'superforce'. The most dramatic thing about supersymmetry (SUSY for short) is that the mathematical tricks needed to change bosons into fermions, and vice versa, automatically, and inevitably, bring in the structure of spacetime, and gravity.

The symmetry operations involved in turning bosons into fermions are close mathematical relatives of the symmetry operations of general relativity, Einstein's theory of gravity. If you apply the supersymmetry transformation to a fermion, you get its partner boson. A quark, say, becomes a squark. Apply the same transformation again, and you get the original fermion back — but displaced slightly to one side. The supersymmetry transformations involve not only bosons and fermions, but also spacetime itself. And general relativity tells us that gravity is simply a reflection of the geometry of spacetime.

But there is a peculiarity about the way physicists came up with the idea of supersymmetry. It all started in 1970, when Yoichiro Nambu, of the University of Chicago, came up with the idea of treating fundamental particles not as points, but as tiny one-dimensional entities, called strings.[1] This was at about the time that the quark model was beginning to be taken seriously, and in the early 1970s Nambu's idea was overshadowed by the rapid acceptance of the quark model — it

---

[1] Historically, the very first hints of string theory came in 1968, when two young researchers at CERN, Gabriel Veneziano and Mahiko Suzuki, were each looking for mathematical functions that could be used to describe the behaviour of strongly interacting particles. They each, independently, noticed that a function written down in the nineteenth century by Leonhard Euler, and called the Euler beta function, might fit the bill. This turns out to be the mathematics underpinning string theory; but it was Nambu who turned the mathematics into physics.

was seen as a rival to quark theory, not as a complementary idea. The fundamental entities that he was trying to model were not quarks, but the hadrons (particles, such as the neutron and proton, which feel the strong force, and which we would now describe as being composed of quarks). The success of the quark model seemed to leave this kind of string theory out in the cold; but a few mathematically inclined physicists played with it anyway.

Nambu's string theory involved spinning and vibrating lengths of string only about 10^{-13} cm long. The properties of the particles he was trying to model in this way (their masses, electric charge and so on) were thought of as corresponding to different states of vibration, like different notes played on a guitar string, or to be attached in some way to the whirling ends of the strings. And these vibrations also involved oscillations in more dimensions than the three of space plus one of time that we are used to — more of this later.

Embarrassingly, though, when the appropriate calculations were first carried through, they said that the entities described by these strings would all have integer spin, in the usual quantum-mechanical sense. That is, they would all be bosons (force carriers, such as photons). And yet, the whole point of the model had been to describe hadrons, which are fermions and have half-integer spin! Then Pierre Ramond, of the University of Florida, found a way around the problem. He found a way of adapting Nambu's equations to include strings with half-integer spin, describing fermions. But those fermionic strings were also allowed by the equations to join together in pairs, making strings with integer spin — bosons. John

Schwarz, in Princeton, Joel Scherk, at Caltech, and the French physicist André Neveu developed this idea into a consistent mathematical theory of spinning strings which included both bosons and fermions, but required the strings to be vibrating in ten dimensions. It was Scherk, in particular, who established, by 1976, that fermions and bosons emerged from this string theory on an exactly equal footing, with every kind of boson having a fermionic partner, and every kind of fermion having a bosonic partner. Supersymmetry had been born.

There is a valuable way of looking at all this, which is often emphasized by Ed Witten, one of the main players in the supersymmetry game in the 1990s. Bosons are entities whose properties can be described by ordinary commuting relationships, familiar everyday rules such as A times B is equal to B times A. Fermions, though, have properties which do not always obey these relationships — they do not commute.[2] The appropriate mathematics that describes this behaviour is quantum mechanics, not classical (Newtonian) mechanics. The concept of fermions is based entirely on the principles of quantum physics, while bosons are essentially classical in nature. Supersymmetry updates our understanding of spacetime to include fermions as well as bosons; it therefore updates the special theory of relativity, Einstein's first theory of space and time, by making it quantum mechanical.

This deep insight was appreciated in 1976, and the next step was seen as being to seek out a way to bring gravity into

---

[2] In fact, they do not commute in a special way; they are said to anticommute.

the fold, updating the General Theory of Relativity, Einstein's second theory of space and time, in the same sort of way. That might have speeded the development of string theory by a decade. But it was not to be — even though the gravity problem was in many people's minds at the end of the 1970s, at that time they saw the next step in terms of an extension of supersymmetry to include gravity, in a theoretical package dubbed supergravity, without using the idea of strings at all.

Almost as soon as supersymmetry had burst upon the scene, the string theory that had given it birth had been forgotten. Never seen as more than a byway of physics by most researchers, it had by 1976 been totally eclipsed by the quark model. Once the idea of supersymmetry had been placed in the minds of physicists, it was easy to incorporate it into the then standard model of the particle world, as we have outlined above. Indeed, that is the way generations of students after 1976 were introduced to supersymmetry, without any mention of strings at all. Physics moved on, and left string theory behind. Just about the only people who carried on working the field were John Schwarz and, over in London, Michael Green (Scherk died young, and made no further contributions to the idea).

But while string theory languished, its offspring, supersymmetry, flourished. A band of enthusiasts soon took up the ideas of SUSY, developing various lines of attack. One describes GUTs in terms of SUSY — the theories are known as SUSY GUTs. Another focuses on gravity — supergravity, which itself comes in various forms with family resemblances but different detailed constructions. One great thing about all the supergravity models is that they each specify a different specific number of possible types of particle in the real world — so many leptons, so many photinos, so many quarks, and so on — instead of the endless proliferation of families allowed by the older GUTs. Nobody has yet succeeded in matching up the specific numbers allowed in any of these supergravity theories with the particles of the real world, but that is seen as a relatively minor problem compared with the previous one of a potentially infinite number of types of particles to worry about. A favoured version of these theories is called 'N = 8' supergravity, and its enthusiasts claim that it could explain everything — forces, matter particles and the geometry of spacetime, in one package. But the best thing about N = 8 supergravity is that it seems not merely to be renormalizable but in a sense to renormalize itself — the infinities that have plagued field theory for half a century cancel out of N = 8 theory all by themselves, without anyone having to lift a finger to encourage them. N = 8 *always* comes up with finite answers to the questions physicists ask of it. 'Superforce' indeed!

But one great puzzle about supergravity is that it requires eleven dimensions in which to operate. Where are they? All this success in the late 1970s and early 1980s in finding potential ways to bring gravity and spacetime back into the fold of particle physics reminded physicists that way back in the 1920s there had already been attempts to explain all the forces of nature in terms of curved spacetime, the way gravity is explained by Einstein's theory. And, from the outset, this approach had not only involved higher dimensions (more than the familiar four), but a neat trick for tucking them away out of sight.

The Many Dimensions of Reality

Early in 1919, Theodor Kaluza, a junior scholar at the University of Königsberg in Germany,[3] was sitting at his desk in his study, working on the implications of the new General Theory of Relativity, which Einstein had first presented four years before and which was about to be confirmed, in spectacular fashion, by Arthur Eddington's observations of light bending during total eclipse of the Sun. As usual, Kaluza's son, Theodor junior, aged nine, was sitting quietly on the floor of the study, playing his own games. Suddenly, Kaluza senior stopped work. He sat still for several seconds, staring at the papers, covered with equations, that he had been working on. Then he whistled softly, slapped both hands down hard on the table, and stood up. After another pause while he gazed at the work on the desk, he began to hum a favourite aria, from Figaro, and started marching about the room, humming to himself all the while.

This was not at all usual behaviour on the part of young Theodor's father, and the image stuck in the boy's mind, so that he was able to recall it vividly sixty-six years later, in an interview for BBC TV's *Horizon* programme.[4] The reason for his father's unusual behaviour was a discovery that is now, after decades in the wilderness, at the heart of research into the nature of the Universe. While tinkering with Einstein's equations in which the gravitational force is explained in terms of the cur-

---

[3] The city is now Kaliningrad, part of the Russian republic.
[4] 'What Einstein Never Knew', first broadcast in 1985.

vature of a four-dimensional continuum of spacetime, Kaluza had wondered, as mathematicians do, how the equations would look if written down to represent five dimensions. He found that this five-dimensional version of General Relativity included gravity, as before, but also another set of field equations, describing another force. The moment that stuck in young Theodor's mind so vividly was the moment when Kaluza senior wrote out those equations and saw that they were familiar — they were, indeed, Maxwell's equations of electromagnetism.

Kaluza had unified gravity and electromagnetism in one package, at the cost of adding in a fifth dimension to the Universe. Electromagnetism seemed to be simply gravity operating in the fifth dimension.

Unfortunately, although Einstein had had no problem in 'finding' four dimensions (three of space and one of time) to put into General Relativity, there was no evidence that there really was a fifth dimension to the Universe. Even so, Kaluza's discovery was striking, and looked important. In those days, a young researcher could not easily publish dramatic new discoveries out of the blue. Today, if you have a bright idea, you write a paper and send it to a learned journal. The journal editors then send it out to an expert (or several experts) to assess, before they decide whether or not to publish. But in those days it was considered correct for the author to send the paper *first* to an eminent authority, who might then, if he approved, send the work on to a learned society with his recommendation that it be published. So Kaluza sent his results to Einstein.

Einstein was initially fascinated and enthusiastic. He wrote to Kaluza, in April 1919, that the idea had never occurred to him, and said 'at first glance I like your theory enormously.'[5] But then he began to pick at little points of detail. A perfectionist himself, he urged Kaluza, in a series of letters, to tidy up these details before publication. The correspondence, and what now seems nit-picking, continued until 1921, when Einstein suddenly had a change of heart (nobody is quite sure why) and sent Kaluza a postcard telling him that he (Einstein) was going to recommend publication. What Einstein recommended in 1921 no journal editor would argue with, and the article duly appeared in the proceedings of the Berlin Academy later that year, under the rather bland title (in German) 'On the Problem of Unification in Physics'.

The obvious defect with the theory presented in that paper (apart from the absence of the fifth dimension) was that it took no account of quantum theory — it was, like General Relativity itself, a 'classical' theory. Even so, Kaluza junior recalls that there was a great deal of initial interest in his father's work in 1922, but then nothing at all. Even Einstein, who spent the rest of his life seeking a unified field theory, seems to have ignored Kaluza's idea from then on, in spite of the fact that in 1926 the Swedish physicist Oskar Klein found a way to incorporate Kaluza's idea into a quantum theory.

The behaviour of an electron, or a photon, or whatever, is described in quantum physics by a set of equations with four variables. A standard form of these equations is called

---

[5] Quoted by Abraham Pais, in *Subtle is the Lord*, page 330.

Schrödinger's equation, after the Austrian physicist who first formulated it. Klein rewrote Schrödinger's equation with five variables instead of four, and showed that the solutions of this equation could now be represented in terms of particle-waves moving under the influence of both gravitational and electromagnetic fields. All theories of this kind, in which fields are represented geometrically in terms of more than four dimensions, are now called Kaluza–Klein theories.[6] As early as 1926, they incorporated gravity and electromagnetism into one quantum theory.

One reason for overlooking, or neglecting, such theories in the years immediately following Klein's work was that there were now more forces to worry about, and so the model seemed unrealistic. The 'answer' is to invoke more dimensions, adding more variables to the equations to include the effects of all the new fields and their carriers, all described by the same geometrical effects as gravity. An electromagnetic wave (a photon) is a ripple in the fifth dimension; the Z, say, might be a ripple in the sixth; and so on. The more fields there are, and the more force carriers, the more dimensions you need. But the numbers are no worse than the numbers that come out of standard approaches to the unification of the four forces, such as supergravity.

---

[6] In fact, Gunnar Nordström, working at what is now Helsinki University, had tried and failed to find a five-dimensional unification of gravity and electromagnetism in 1914, and in 1926 H. Mandel independently came up with the same basic idea as Kaluza, apparently in ignorance of Kaluza's 1921 paper.

Indeed, the numbers are *exactly the same*. The front runner among supergravity candidates (indeed, the only good supergravity theory) is the N = 8 theory. That theory describes a way to relate particles with different spins, under the operations of supersymmetry. The range of spins available is from +2 to −2, and spin comes in half-integer quanta.[7] So there are eight steps (eight SUSY transformations) involved in getting from one extreme to the other, hence the name. But there is another way of looking at all this. Just as Kaluza tinkered with Einstein's equations to see how they would look in five dimensions, so modern mathematical physicists have tinkered with supergravity to see how it would look in different dimensions. It turns out that the simplest version of supergravity, the most beautiful and straightforward mathematical description, involves eleven dimensions — no more, no less. In eleven dimensions, there is a unique theory which just might be the sought-after superforce. If there are eleven dimensions to play with, all the complexity of the eight SUSY transformations disappears, and we are left with just one fundamental symmetry, an N = 1 supergravity. And how many dimensions does Kaluza–Klein theory need to accommodate all of the known forces of nature and their fields? Precisely eleven: the four familiar components of spacetime and seven additional dimensions — no more, no less.

The implications of all this have excited many physicists,

---

[7] The hypothetical 'particle' of gravity, the graviton, has spin 2, and theory suggests that this is the greatest value possible.

and no less an authority than Abdus Salam described this geometrization of the world of particles and fields as 'an incredible, miraculous idea'.[8] They are still a long way from producing a fully worked out theory of this kind, but the unification of Kaluza–Klein theories with supergravity to produce an eleven-dimensional supergravity theory was a key development in the search for SUSY — although its true significance is only apparent with hindsight.

But still, why don't we 'see' all those extra dimensions? This is no problem to the mathematicians. Each of those extra dimensions must have, somehow, become curled in upon itself, becoming invisible in our three- (or four-) dimensional world. The analogy that is often made is with a hosepipe. From a distance, a hosepipe looks like a wiggly line, a one-dimensional object. But closer up you can see that it is a cylinder, a two-dimensional object. Every 'point' on the wiggly line is actually a circle, a loop around the point, and the cylinder is the string of all these circles, one behind the other. In the early Kaluza–Klein theory, every point of spacetime is thought of as a little loop — a loop just 10^{-32} cm across, extending in a direction that is neither up, nor down, nor sideways. The modern version is a little more complicated. The 'loop' becomes an object called a seven-sphere (actually a slightly squashed seven-sphere), the seven-dimensional analogue of a sphere. But the principle is the same. And, mathematicians tell us, the seven-sphere is the *simplest* form of

[8] Horizon, *op. cit.*; see also Salam's paper with J. Strathdee, in *Annals of Physics*, volume 141, page 316.

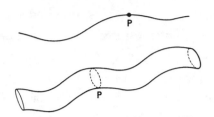

Figure 4.2 What seems from a distance to be a one-dimensional line turns out here to be a two-dimensional tube. Each 'point' on the line is really a little circle going round the circumference of the tube. This is how compactification is thought to work to 'hide' the extra dimensions of space required by some theories of particle physics.

multi-dimensional structure that still allows for the Universe we see around us to be as complex as it is.

On this picture, the Universe was born in an eleven-dimensional state, and there was no distinction between force and matter, let alone between different kinds of force, just some kind of pure state of eleven-dimensional energy. As the energy dissipated, some of the dimensions curled up upon themselves, creating the structures that we think of as matter — the 'particles' — as waves vibrating in the coiled-up dimensions, and creating the forces of nature as the visible manifestation of the distortions of the underlying geometry. To crack the seven-sphere and unpeel it to reveal the ten dimensions of space in all their glory would require more than even the grand unification energy. It requires, indeed, the energy of creation itself.

It is all heady stuff, at the cutting edge of current research, and new ideas eddy around in profusion in the scientific journals of the 1990s. The most exciting variation on the theme, as I shall shortly explain, treats 'particles' not as points but as one-dimensional strings which 'move' in a ten-dimensional spacetime. These are 'superstring' theories. On the other hand, some theorists, including, surprisingly, Stephen Hawking (who was one of the first big fans of $N = 8$ theory, and said that it might mark the end of physics by explaining everything that physicists ever set out to explain), see all of the Kaluza–Klein ideas as leading up a blind alley.

The theorists who like the Kaluza–Klein version of supergravity and SUSY today like it, not because of any experimental proof that it is correct, but because it is so beautiful and internally consistent. As Einstein once said of General Relativity, it is so beautiful that it *must* be true! Kaluza himself would have appreciated the point, for he was a theorist *par excellence*. His son tells how Theodor Kaluza senior taught himself to swim by reading a book. Having thoroughly absorbed the theory, and convinced himself that it was true, he took the family off to a nearby lake, jumped in and proceeded to swim, fifty metres out and fifty metres back. He proved that the theory worked. We don't have a suitable lake in which to throw the Kaluza–Klein theory to see if it will sink or swim. Like Salam, I can only say that I would like it to be correct.

But there is one potentially devastating snag. In order to accommodate spin into such a theory, it turns out that overall space and time together must occupy an even number of dimensions. As you may have noticed, eleven is an odd number. But just as theorists realized the problems that this might pose, a new variation on the idea of superforce emerged, taking on board both the idea of supersymmetry and the notion of higher dimensions, and more besides.

Stringing Things Together

The search for a unified theory of physics can be thought of in terms of the two great theories of twentieth-century physics. The first, the general theory of relativity, relates gravity to the structure of space and time. It tells us that they should be treated as a unified whole, spacetime, and that distortions in the geometry of spacetime are responsible for what we perceive as the force of gravity. The second, quantum mechanics, describes the behaviour of the atomic and subatomic world; there are quantum theories which describe each of the other three forces of nature, apart from gravity. A fully unified description of the Universe and all it contains (a 'theory of everything', or TOE) would have to take gravity and spacetime into the quantum fold. That implies that spacetime itself must be, on an appropriate very short-range scale, quantized into discrete lumps, not smoothly continuous. String theory, in an extended form known as superstring theory, *naturally* produces a description of gravity out of a package initially set up in quantum terms, but it took several years for gravity to fall out of the equations.

String theory only took off in the middle of the 1980s, after a new variation on the theme was developed by John Schwarz and Michael Green. They started working together at the end of the 1970s, after meeting at a conference at

CERN, and discovering that, unlike everybody else studying particle physics at the time, they were both interested in string. They began to produce results almost immediately. The key first step that they took was to realize that what was needed was indeed a theory of *everything* — all the particles and fields — not just hadrons. In such a theory, strings would have to be very small indeed — much smaller than Nambu's strings, which were only ever designed to describe hadrons. Even without knowing how the theory would develop, Schwarz and Green could predict what scale the strings would operate on, because they wanted to include gravity in the package. Gravity becomes seriously affected by quantum effects at a scale around 10^{-33} cm (that is, 10^{-35} m), the distance scale at which the very structure of spacetime itself becomes affected by quantum uncertainty.[9] And, of course, from the outset the new version of string theory had SUSY built into it.

The first string model to be developed by Schwarz and Green, in 1980, dealt with open-ended strings vibrating in ten dimensions, able to link up with one another and break apart. Superficially, it doesn't look like anything more than a shrunken version of Nambu's string theory. In fact, it went far further, including (in principle — actually carrying the calculations through was another matter) string states correspond-

---

[9] The scale at which quantum effects become important for a particular force depends on the strength of the force. Quantum effects only become dominant for gravity on such a tiny scale because gravity is such a weak force, by far the weakest of the four forces of nature. It is by measuring the strength of gravity (the value of the gravitational constant, G) that physicists can work out what scale quantum gravity operates on.

ing to all the known particles and fields, and all the known symmetries affecting fermions and bosons, plus supersymmetry. Or rather, I should say 'almost all the known particles and fields'. There was one exception — gravity. In spite of their good intentions, gravity still could not be explained by the new string theory.

Nevertheless, this early version set the scene for what was to follow. The central idea of all subsequent string theories is that the conventional picture of fundamental particles (leptons and quarks) as points with no extension in any direction is replaced by the idea of particles as objects which have extension in one dimension, like a line drawn on a piece of paper, or the thinnest of strings. The extension is very small — about 10^{-35} of a metre. It would take 10^{20} such strings, laid end to end, to stretch across the diameter of a proton.

The next big step towards a true theory of everything

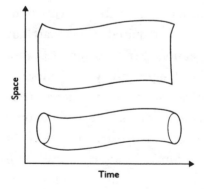

Figure 4.3 An open string traces out a sheet as it moves through spacetime. A loop of string traces out a tube.

came in 1981, when Schwarz and Green introduced a new twist (literally) to the story. The open string theory of 1980 became known as the Type I theory, and the new, Type II theory introduced a key variation on the theme — closed loops of string. Type I theory only had open-ended strings; Type II theory only had closed loops of string. In a particularly neat piece of packaging, in closed loops fermionic states correspond to ripples running around the loop one way, while bosonic states correspond to ripples running around the loop the other way, demonstrating the power and influence of supersymmetry. The closed loop version had some advantages over the open model, not least because it proved easier to deal with those infinities that plague particle physicists in the closed loop model. But the Type II theory also had its difficulties, and did not seem (at the time) capable of predicting, or encompassing, all the variety of the known particle world.

There was one other cloud on the horizon. In 1982, Ed Witten and Luis Alvarez-Gaumé discovered that the Kaluza–Klein compactification trick will only work to make the forces of nature in the way required if you start out with an odd number of dimensions before compactification. This made eleven-dimensional supergravity look more attractive than ever before, but gave the ten-dimensional string theories real problems. This didn't stop people working on those theories, but it gave them something extra to think about.

The next step forward was actually a step back. Dissatisfied with Type II theory, Schwarz and Green went back to Type I theory, and tried to remove the infinities which plagued it. The problem was that there were many possible variations on their theme, and that all of them seemed to be beset not just by infinities but by what were called anomalies — behaviour which did not match the behaviour of the everyday world, especially its conservation laws. For example, in more than one version of the theory, electric charge is not conserved, so charge can appear out of nothing at all, and disappear, as well.

But in 1984 Schwarz and Green found that there is one, and only one, form of symmetry (technically, SO(32)) which, when applied to the Type I string theory, removes all of the anomalies and all of the infinities. They had a unique theory, free from anomalies and infinities, that was a real candidate for the theory of everything. It was at this point that other physicists started to sit up and take notice of strings once again.

One of the teams that was fired up by the success achieved by Green and Schwarz in 1984 was based at Princeton Uni-

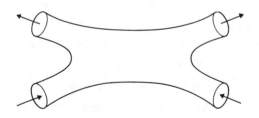

Figure 4.4 The interaction between two loops of string can be represented in a Feynman diagram. Crucially, though, because the strings trace out tubes through spacetime, none of the interactions occur at a point in such a diagram; everything is blurred out slightly. It is because none of the interactions occur at a point that infinities do not arise in string theory.

versity. David Gross and three colleagues (together sometimes known as 'the Princeton String Quartet') took a second look at the closed loop idea, writing it down using a different mathematical approach. There was plenty for them to write down, because the theory is a little more complicated than I have let on so far. The kind of vibrations associated with fermions do indeed require the ten dimensions that I have mentioned. But the bosonic vibrations described (at first, unintentionally) by Nambu's first version of string theory actually take place in twenty-six dimensions. Gross and his colleagues found a way to incorporate both kinds of vibration into a single closed loop of string, with the ten-dimensional vibrations running one way round the loop and the twenty-six-dimensional vibrations running the other way around the loop. This version of the idea is called the heterotic string theory.[10]

The heterotic strings neatly tidy up a loose end in the Type II theory. In theories involving open strings, some of the properties we associate with particles (the properties which physicists call 'charges') are tied to the end points of the whirling strings (this may be electric charge, if we are dealing with electromagnetism, or the 'colour charge' of quarks, or something else). But closed loops do not have end points so where are these properties located? In heterotic strings, these properties are still properly described, but have to be thought of as somehow smeared out around the strings. This is the main physical distinction between heterotic strings and the

---

[10] 'Heterotic' from the same Greek root as in 'heterosexual', implying a combination of at least two different things.

kind of closed strings that Green and Schwarz investigated at the beginning of the 1980s; you can picture heterotic string theory as a kind of hybrid combination of the oldest kind of string theory and the first superstring theory.

How can two different sets of dimensions apply to vibrations of the same string? Because for the bosonic vibrations sixteen of the twenty-six dimensions have been compactified as a set, leaving ten more which are the same as for the ten-dimensional fermionic vibrations, with six of those ten dimensions being compactified in a different way to leave us with the familiar four dimensions of spacetime. It is the extra richness provided by the sixteen extra dimensions that makes for the richness in the variety of bosons, from photons to W and Z particles and gluons, compared with the relative simplicity of the fermionic world, built up from a few quarks and leptons. The sixteen extra dimensions in the heterotic string theory are responsible for a pair of underlying symmetries, either of which can be used to investigate the physical implications of the theory (any other choice of symmetry groups leads to infinities). One of these is the $SO(32)$ symmetry group, which had already turned up in the investigation of open strings (32, of course, being twice 16); the other is a symmetry group known as $E_8 \times E_8$, which actually describes two complete worlds, living alongside each other (8 plus 8 also being 16). Each of the E_8 symmetries can be naturally broken down into just the kind of symmetries used by particle physicists to describe our world. When six of the ten dimensions involved curl up, they provide a symmetry group known as E_6, which is itself broken down into $SU(3) \times SU(2) \times U(1)$. But

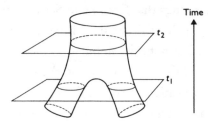

Figure 4.5 Loops of string can perform another trick — two loops can merge with one another to make a third loop, as in this 'trouser' diagram.

SU(3) is the symmetry group associated with the standard model of quarks and gluons, while SU(2) × U(1) is the symmetry group associated with the electroweak interaction. Everything in particle physics is included within one of the E_8 parts of the overall $E_8 × E_8$ symmetry group.

Since only one of the E_8 components is needed to describe everything in our Universe, that leaves a complete duplicate set of possibilities. The symmetry between the two halves of the group would have been broken at the birth of the Universe, when gravity split apart from the other forces of nature. The result would, some theorists believe, be the development of two universes, interpenetrating one another but interacting only through gravity — our world and a so-called 'shadow' universe. There would be shadow photons, shadow atoms, perhaps even shadow stars and shadow planets,[11] inhabited by shadow people co-existing in the same spacetime that we in-

---

[11] And, indeed, shadow sleptons, shadow squarks and shadow bosinos; SUSY itself would be duplicated in the shadow world.

habit, but forever invisible. A shadow planet could pass right through the Earth and never affect us, except through its gravitational pull. It sounds like science fiction, but one reason that the idea has been taken seriously is that there is astronomical and cosmological evidence that a lot of the Universe exists in the form of dark matter, detectable gravitationally but not seen. It is, though, at least as likely that in the shadow universe later symmetry breakings occurred differently from in our own world, so that there are no shadow stars, and so on, after all. All this is tangential to the story being told here (but see *In Search of the Big Bang*). It has, though, brought us back to gravity, and gravity is the reason why interest in string theory and supersymmetry exploded in the middle of the 1980s.

The excitement was largely to do with the way gravity appeared naturally out of superstring theory. You can think of gravity in two ways. Starting out from Einstein's description of curved spacetime, you are led to the image of gravity waves, ripples in the fabric of spacetime, with, of course, an associated particle, the spin-2 graviton. This was the way the concept appeared historically. But you could, in principle, start out with a quantum field theory based on a zero mass, spin-2 particle, the graviton, and see what the equations describe. Carry the calculations through, and you end up with Einstein's general theory of relativity.[12] The problem with all

---

[12] This was first spelled out by Richard Feynman in the 1960s, in a series of lectures he gave for graduate students at Caltech; his lectures on gravity were only published in the mid-1990s, as *Feynman Lectures on Gravity*.

theories prior to superstrings (except, just possibly, $N = 8$ supergravity) is that when you add in a massless, spin-2 particle they are beset with infinities that are impossible to remove even by renormalization. The dramatic discovery that emerged in the mid-1980s is that whenever theorists set up a mathematical description of superstring loops, tailored to describe the behaviour of the known particles, a description of a massless, spin-2 particle *always* falls out of the equations along with the quarks and leptons and the rest. What's more, it does so without causing those uncomfortable infinities to rear their ugly heads. One of the founders of superstring theory, John Schwarz, refers to this as a 'Deep Truth', that must be telling us something fundamental about the way the Universe works.

Gravity *must* be included in superstring theory, and arises naturally in a way that can be portrayed in simple physical terms. The simplest form of closed string that emerges automatically from the theory has the properties of a spin-2 vector boson, the quantum particle of gravity. Indeed, they *are* gravitons, the carriers of the gravitational force. Gravity, including Einstein's equations of the general theory of relativity, emerges naturally from string theory as a quantum phenomenon.

All this was exciting enough to encourage more theorists to begin working on strings and superstrings after 1984. But, just as there had been a gap of about ten years between the first breakthrough in string theory (involving SUSY itself) in the mid-1970s and the combination of the discovery of heterotic strings and that gravitons are a part of string theory in the mid-1980s, so it was to be another ten years before the next breakthrough occurred. It might have happened sooner — remember that puzzle about the need for an odd number of dimensions to make compactification work? And the joker in the pack, eleven-dimensional supergravity? In the late 1980s, a few theorists, including Michael Duff, of Texas A & M University, raised the possibility that we ought not to be dealing with strings at all, but ought to add in another dimension, making them resemble two-dimensional sheets (membranes) rather than one-dimensional lines. The extra dimension brings the total number up to eleven, but one of these dimensions is immediately rolled up so that the membrane behaves like the ten-dimensional string of string theory. The idea was more a speculation than a fully worked out theory, and it was laughed out of court at the end of the 1980s. But it was revived, as a much more complete theory, in the mid-1990s, and today the membrane idea (often referred to as M-theory[13]) is just about the hottest game in town.

The reason why M-theory is causing excitement at the end of the 1990s is that it offers, at last, a *unique* mathematical package to describe all of the forces and particles of nature. As you have noticed, string theory itself comes in several different varieties, which each have their good points and their bad points. In fact, there are exactly five variations on the theme. These are the Type I theory of Schwarz and Green, two versions of the Type II theory, and the two versions of heterotic strings. In addition, there is still that wild card, the eleven-dimensional

---

[13] The term was coined by John Schwarz, who says that the 'M' can stand for 'magic, mystery or membrane, according to taste'.

supergravity. It can be shown mathematically that these are the only viable variations on the theme; all of the other possibilities involving supersymmetry are plagued by infinities.

Now, at first sight six rival contenders for the title 'theory of everything' looks a lot. But, in fact, this is a remarkably short list. The old-fashioned particle physics approach to grand unified theories gives you a plethora of possibilities, any of which are just as good as any of the others. To have only half a dozen theories to choose from seemed miraculous in the 1980s. The dramatic new discovery made in the mid-1990s, however, was that all six theories are related to one another. Specifically, they are all different manifestations of a single M-theory. In a manner reminiscent of the way that the electroweak theory is a single theory that describes what seem to be two separate interactions at lower energies (electromagnetism and the weak interaction), M-theory is a single theory at even higher energies, and describes what seem to be six different models at lower energies. Specifically, the differences between the six models appear at the level of the weak interaction, and the unity is clear at the level of the strong interaction.

We may not have to wait too long to find out just how good a theory M-theory really is, and whether it is indeed the long sought-after theory of everything. The kind of energies needed to probe the predictions of M-theory should be achieved at the latest high energy particle accelerator, the Large Hadron Collider (LHC), which is expected to begin operating at CERN in the middle of the first decade of the twenty-first century. This follows, coincidentally, the pattern in which major developments in string theory have occurred in the middle of each of three successive decades — the 1970s, the 1980s, and the 1990s. It would certainly be neat if the story were rounded off in 2005 and 2006, thirty years after the first revolution in string theory. The ultimate test of any theory, of course, is by comparing its predictions with the results of experiments.

Meanwhile, mathematicians are very much involved in the new physics. Movement of points through space and time can be described in terms of lines traced out by the particles — trajectories, or world lines. Moving strings and membranes, however, trace out sheets and volumes in spacetime, which require a quite different mathematical treatment. Multidimensional topologies, which some mathematicians have studied for their intrinsic abstract interest, suddenly turn out to have practical relevance. There is, it seems, something for everybody in superstring theory.

But we may not have to wait even another five or six years to have confirmation of the existence of SUSY itself. After all, identifying just one SUSY particle in the lab would prove that supersymmetry is a good description of the world, and would hint strongly that some form of superstring theory, perhaps M-theory, is the underlying theory of everything.[14] There is already some, albeit indirect, good experimental evidence for SUSY. If we take one step back from the hopes of ultimate uni-

---

[14] There have, indeed, been a couple of claims already that SUSY particles (possibly selectrons) have been detected. But these each correspond to fleeting 'one-off' events in particle accelerators that nobody has, as yet, been able to replicate. By the time you read this book, though, there may have been some more such sightings.

fication of all four forces, including gravity, into one package with the help of string, we find that in 1991, as well as confirming the accuracy of QCD, the Large Electron-Positron Collider at CERN in Geneva was also producing new evidence that the basic concept of supersymmetry, which underpins both superstring theory and other attempts to come up with a theory of everything, really does provide us with a good description of the fundamental behaviour of particles and fields.

SUSY Found?

The most effective way to test the predictions of supersymmetry would be to actually create the supersymmetric partners to at least some of the everyday particles in high energy collisions at accelerators like LEP, and to measure their properties. Since no SUSY particle has yet been produced in this way, we know that they have masses (assuming they really do exist) corresponding to energies greater than those attainable in the present generation of particle accelerators. That means masses greater than a few hundred GeV (1 GeV, roughly the mass of a proton, is 1.58×10^{-27} of a kilogram). But there is a more subtle way to test for the existence of supersymmetry.

This trick starts out from the assumption that the forces of nature — sometimes known as the 'interactions' — really are unified at some very high energy. Then, existing measurements of the way particles interact can be used to extrapolate how their behaviour will change as the energy available increases. Finally, theorists hope, it should be possible to show that these extrapolations only point towards grand unification if we make allowance for the effects of supersymmetry. The exact allowance for the SUSY influence required to ensure unification would then give us an insight into the properties of the SUSY particles themselves.

This is the technique that was applied at CERN. It uses measurements of a basic property of any quantum field which is known as the coupling constant. The coupling constant is in effect a number (a 'pure' number, without any dimensions attached to it, unlike length or mass) which determines the strength of a particular interaction. It is by comparing coupling constants that we are able to say that the weak interaction is a certain number of times weaker than the strong interaction, for example. But when I gave some numbers for such comparisons between the four forces earlier, I didn't mention that in fact the numbers I quoted all apply at low energies, and the strength of the relevant coupling constant in each case depends on the energy involved in the interaction between two particles. For example, the electromagnetic coupling constant which helps to describe the way in which two electrons scatter from each other at low energies is $1/137$ — a number that may be familiar, if you have studied physics, as the fine structure constant. But in collisions at LEP, which occur at energies of around 100 GeV, the coupling is stronger, and the relevant constant is $1/129$. This increase in coupling strength at higher energies is predicted by the standard model of particle physics, and the measured value for interactions at LEP is a strong piece of evidence that the standard model is very much on the right lines. But the standard model does not predict the precise value of the coupling constant (indeed, the reasons for the precise values of all such constants of nature remain a deep mystery which no theory can yet explain).

There is another coupling constant for the strong interaction, which also varies as the energy of interactions (this time, interactions involving quarks and gluons, rather than electrons and photons) increases, but in the opposite sense. At higher energies, the strong coupling constant is smaller than at low energies. Again, observations match the broad pattern predicted by theory. Measurements of this constant at low energies give a value of about 0.18, while at LEP energies the measured value is 0.12. This weakening of the strong coupling at high energies is directly related to the way in which quarks are held together more loosely when they are close together, and more strongly when you try to pull them apart. The third force involved in the standard model, the weak interaction, also has a coupling constant that gets smaller at higher energies. But here the plot begins to thicken.

The carriers of the weak interaction, the W and Z bosons, have, as we have seen, large masses — 80 GeV for the Ws and 91 GeV for the Z. Because of this, the weakening of the relevant coupling constant only shows up at energies greater than those equivalent to the masses of the W and Z — that is, at around 100 GeV, where LEP operates. And as a further complication, because the electromagnetic and the weak interactions are separate facets of the unified electroweak interaction, it turns out that it is easier to do the comparisons by expressing them in terms of two combinations of the weak and electromagnetic coupling constants. These effective coupling constants are labelled α_1 and α_2. α_1 increases as the energy involved in interactions increases, while α_2 decreases as the energy involved increases. In the same notation, the strength of the coupling constant of the strong interaction is labelled α_3.

Using data from LEP, researchers at CERN were able, in 1991, to measure the variation in these three parameters over a range of energies. They could then plot these variations as three lines on the same graph, and extend each of the straight lines they found out to much higher energies, where no experiments have yet been carried out. If grand unification of the three interactions does take place at some higher energy, then the three lines ought to cross at a single point corresponding to that energy. But they do not — in fact, the three lines cross one another to form a tiny triangle on the graph, at an energy of about 10^{16} GeV.[15]

So what happens if we add SUSY into the calculations? If the SUSY particles all have much the same mass, and their average mass is bigger than the mass of the Z particle, then the extrapolation of the changes in the coupling constants is just the same as before from an energy corresponding to the Z mass up to an energy corresponding to the SUSY mass. But the graphs then bend, at an energy corresponding to the SUSY mass, before continuing once more as straight lines out into the region of very high energies where (we hope) grand unification occurs. By choosing a specific value of the average

---

[15] Note that in a further piece of confusion for the non-specialists, what the experts actually plot on these graphs is the *inverse* of ('one over') the coupling constants. So the inverse strong constant, for example, gets *bigger* at higher energies.

SUSY mass, the CERN team can make the three lines cross at a point; the value of the SUSY mass needed to do this trick is only a little higher than the energies so far reached in particle colliders, at around 1,000 GeV. This is excellent news, since the next generation of particle accelerator, the Large Hadron Collider at CERN, will be able to probe precisely this energy range, testing the theory of supersymmetry (and doing some other neat tricks discussed in Appendix II). If all the calculations are correct, and SUSY particles do exist, then they may be discovered before the year 2010. And if those accelerators fail to find SUSY particles with masses around 1,000 GeV? Well, then it will be back to the drawing board to find some other way to make the three lines meet at a point. We are that close to finding out, at last, just how good SUSY really is.

I'm certainly optimistic. The bottom line, so far, is that at the very least adding SUSY in to the calculations brings the extrapolation of the coupling constants to high energy closer to unification than the extrapolation using the standard model without SUSY, which is already a significant step. It is even more remarkable that this has been achieved using the simplest possible extension of the standard model to include SUSY, and with the simplest possible assumptions about grand unification (although all of these simpler models fall naturally out of superstring theory). I suspect that we are indeed in the presence of another Deep Truth, and that is indeed a suitably upbeat note on which to end my story of the search for SUSY.

Appendix I

Group Theory
for Beginners

Group theory is a branch of mathematics that deals with groups and symmetry. In mathematics, a group (or symmetry group) is defined as a collection of elements (a set), which are labelled a, b, c, and so on, *and* which are related to one another by certain rules:

First, if a and b are both members of the group G, then their product, ab, is also a member of the group G. This process is associative, which means that $a(bc) = (ab)c$, and so on.

Second, there must be an element, called the unit element and usually denoted by e, defined so that $ae = a$, $be = b$, and so on for all elements in the group.

Third, each element has an inverse, written as a^{-1}, b^{-1}, and so on, defined so that $aa^{-1} = e$, and so on.

A group for which $ab = ba$ is an Abelian group. The set of ordinary integer numbers (1, 2, 3, . . .) is a simple example of an Abelian group. More generally, groups are made up of elements which are themselves matrices. A matrix is a kind of two-dimensional number, represented by ordinary numbers laid out in a grid, like pieces on a chess board. If the smallest object that represents a particular group is a matrix made up of N rows and N columns (an $N \times N$ matrix), then N is the dimension of the group. This is where the number 3 in the group SU(3), which turns out to be important in particle theory, comes from — it is the dimension of that particular group (the 'SU' stands for 'special unitary' group).

Group theory was developed in the nineteenth century by the Norwegian mathematician Sophus Lie (so these groups are sometimes known as 'Lie groups'). Although the theory had been used in mathematical descriptions of the symmetry of crystals, it was a largely obscure branch of mathematics until the second half of the twentieth century, when Chen Ning Yang and Robert Mills found a way to describe the strong interaction in terms of Lie groups, and then Murray Gell-Mann and Yuval Ne'eman (working independently of one another) found that SU(3) provided a framework for describing mathematically the relationships between elementary particles. Since then, symmetry groups have been an essential tool used by physicists in their development of gauge theories of the forces of nature. In this context, symmetry groups are sometimes called gauge groups.

A simple example of a group is the set of rotations of a coordinate system (an x, y graph system) around the point where the x and y axes meet. If you turn the axes of the graph around, the coordinates of every point measured relative to those axes changes, but the relationships between those points does not change — it is simply a relabelling exercise, a gauge transformation. This means, for example, that although the Earth is rotating, the distance between London and Paris (or between any other two points on the globe) stays the same. We are all experiencing a gauge transformation, literally every minute of the day.

And if you turn the axes first through an angle A and then through an angle B, the effect is just the same as turning them through the angle C, where C = A + B. Because the angle of rotation can be as small as you like and varies smoothly, as in the example of the rotating Earth, this rotation group is called a continuous group (the SU groups that are so important in particle theory are also continuous groups). The fact that the laws of physics are unchanged by such a rotation implies the law of conservation of angular momentum; in general, whenever a symmetry group describes the behaviour of a physical phenomenon, there must be some conserved quantity associated with that phenomenon (this is sometimes referred to as Noether's theorem, and is a useful feature of group theory which can be used to provide physical insights into the behaviour of particles and forces).

The groups describing the behaviour of particles and fields in the quantum world are, alas, harder to visualize in physical terms, but obey the same mathematical principles. One of the key features of this application of group theory is that because of the inherent symmetries they predict that there should be certain numbers of particles of a particular type (quarks, say, or gluons), described by a particular symmetry group. SU(3), for example, has 'room' for quarks with just three different varieties of colour charge, and for just eight different varieties of gluon.

One of the key features of the kind of groups important in particle physics theory is their symmetry. We all know what symmetry is in the context of a geometrical pattern, and this idea is carried over into the quantum world to describe the relationships between forces and particles. This enables scientists to describe physics in terms of geometry — if necessary, invoking more dimensions than the familiar three of space plus one of time.

A familiar example of symmetry is the reflection symmetry of some patterns, in which the right-hand side of the pattern is a mirror image of the left-hand side. Another symmetry is possessed by a perfect sphere — it looks the same no matter how it is rotated, and is said to possess spherical symmetry, or to be rotationally invariant.

Symmetry is built into the laws of nature in a very deep way. The symmetry which says that the laws of nature are the same at every place in the Universe (translational invariance), for example, corresponds to the law of conservation of linear momentum; the symmetry which says that the laws of physics are the same at all times is equivalent to the law of conservation of energy; and the rotational invariance of the laws of physics is equivalent to the law of conservation of angular momentum.

Many of the symmetries in quantum physics are broken symmetries, where a situation that is intrinsically symmetrical has become asymmetrical — the classic example is of a ball balanced on a perfectly cone-shaped hill, which is a symmetrical situation; when the ball rolls off down one side of the hill, the symmetry is broken, but the situation you end up with still carries some imprint of the underlying symmetry. Using these ideas, physicists have discovered symmetries between the forces of nature, between quarks and leptons, and even between fermions and bosons (supersymmetry).

One specific kind of symmetry is at the heart of field theory. This is gauge symmetry, a concept used in field theory to describe a field for which the equations describing the field do not change when some operation is applied to all particles everywhere in space. (It is also possible to have local symmetry, where the operation is applied in some particular region.)

The term 'gauge' simply means 'measure', and the point is

that fields with gauge symmetry can be regauged (or remeasured) from different baselines without affecting their properties.

The classic example is gravity. Imagine a ball sitting on a step on a staircase. It has a certain amount of gravitational potential energy. If the ball moves down to another step on the staircase, it loses a specific amount of gravitational energy, which depends only on the strength of the gravitational field and the difference in height of the two steps. You can measure the gravitational potential energy from anywhere you like. It is usual to measure from either the surface of the Earth or the centre of the Earth as your baseline, but you could choose any of the steps, or any point anywhere in the Universe as the zero for your measurements. The point is that the difference in energy between the two steps will always be the same, no matter how you regauge your baseline. So gravity is a gauge theory.

Gravity and electromagnetism are both gauge theories, and the requirement of gauge symmetry was one of the key inputs used in the development of the theory of the weak interaction and the theory of quantum chromodynamics in terms of quantum fields. The situation is more complicated in these quantum field theories than in the simple example of gravity (see gauge symmetry), but can be pictured with the help of an analogy developed by Heinz Pagels in his book *The Cosmic Code*.

Pagels asks us to imagine an infinite sheet of paper painted a uniform shade of grey. It is completely uniform, and there is no way to tell where you are on the sheet of paper — it is globally invariant. The same is true whatever the exact shade of grey of the paint, an example of gauge symmetry ('regauging' the colour makes no difference). Now imagine a similar sheet of paper painted in different shades of grey. The symmetry is broken and it is easy to tell different regions of the sheet from one another. But the symmetry — the global invariance — can be restored if we lay over the multishaded paper a sheet of clear plastic which has been painted in shades which exactly balance the pattern in the paper — dark where the paper is light, light where the paper is dark. The combined effect will be to produce a uniform shade of grey, with the global invariance restored.

The sheet of multi-shaded paper represents a visible quantum field. The sheet of plastic painted in complementary shading represents a gauge field, sometimes called a Yang–Mills field after the two researchers who developed this approach to quantum field theory in the 1950s, which restores the symmetry.

The key point is that a completely globally invariant field is undetectable, because it is the same everywhere. Fields only express themselves, as it were, when the symmetry is broken and there are differences from place to place. It was this idea of broken symmetry in gauge theory that led Steven Weinberg and Abdus Salam (working independently of one another) to develop the basis of the electroweak theory in 1967, pointing the way for all subsequent attempts to develop a grand unified theory.

Appendix II

Recreating the Birth of the Universe in the Lab

Theories of the fireball in which the Universe was born may soon be tested by experiments on Earth which recreate the conditions of the Big Bang itself in a series of 'little bangs'. If the Universe was born out of a hot fireball of energy, as the widely accepted Big Bang theory tells us, how did that energy get converted into the matter that we see around us today? The standard theory of matter says that ordinary hadrons — particles such as the protons and neutrons that make up atomic nuclei — are composed of fundamental entities known as quarks, which are held together by swapping gluons between themselves. The exchange of gluons produces a force so strong that no individual quark can ever escape from a hadron. But under the conditions of extreme pressure and temperature during the first split second of the Universe, 13 billion years ago, individual hadrons could not have existed. Instead, according to standard theory, the Universe consisted of a soup of quarks and gluons — a 'quark-gluon plasma'.

The quark-gluon 'era' ended about one hundred-thousandth

of a second after the Universe began expanding from a point. At this critical time, a phase transition took place, equivalent to the way steam changes into liquid water, and hadrons were formed. Physicists on both sides of the Atlantic are now planning experiments which will probe the quark/hadron transition, providing experimental tests of the theories on which our understanding of the early Universe is based.

To get a feel for just how extreme the conditions involved are, we need to look at temperature and density in terms rather different from those of everyday life. Confusingly, physicists measure both quantities in terms of what seems to be the same unit — the electron volt, or eV. Strictly speaking, this is a measure of energy, with 1 eV equivalent to 1.6×10^{-5} joule, so it is a perfectly good measure of temperature. Particles which collide with one another with kinetic energies of a few electron volts have a temperature equivalent to a few tens of thousand degrees, on the Kelvin scale, and these are the energies and temperatures associated with ordinary chemical reactions.

Energy can be converted into a mass equivalent by dividing by c^2, in line with Einstein's famous equation $E = mc^2$, and when electron volts are used as units of mass, or in expressing densities, the division by c^2 is taken as read. In those terms, the mass of an electron is 500 keV, and the mass of a proton is 1 GeV. The neutron has almost the same mass (actually slightly more), and the packing of neutrons and protons together in atomic nuclei provides the greatest density of matter which can exist in the Universe today (except for a faint possibility that hadrons may be squeezed so hard in the centres of some neutron stars that they are squashed into a quark-gluon soup). The radius of a proton is about 8×10^{-16} m, which is near enough, for our present needs, to one femtometre (1 fm = 10^{-15} m). So the density of a proton —

the ultimate density of everyday matter — is, in round terms, 1 GeV per fm^3.

Calculations of the way in which matter behaves at the quark-hadron transition have been carried out using a relatively new technique called lattice gauge theory, using powerful computers. The critical temperature (corresponding to the critical temperature at which water boils) is in the range from 150 to 200 MeV, according to these calculations, and this corresponds to an *energy* density of 2–3 GeV per fm^3 — that is, enough pure energy present in the volume of a single proton to create *three* protons in line with Einstein's equation. How can physicists set about creating such extreme conditions?

The line of attack which is now being followed at CERN, in Europe, and at the Brookhaven National Laboratory, in the United States, is to collide beams of heavy ions head on. Particle accelerators routinely carry out experiments in which beams of protons or electrons (or their antimatter counterparts) are smashed into targets containing nuclei of heavier elements, or into opposing beams of elementary particles. But now researchers are developing the technique required to take beams containing nuclei of very heavy elements and smash them into opposing beams containing the same type of heavy nuclei. To get a picture of the kind of collision that will result when two such nuclei meet head on, consider the (as yet hypothetical) example of what happens to a gold nucleus accelerated to 0.999957 times the speed of light.

A gold nucleus contains 118 neutrons and 79 protons; so it has 79 units of positive charge, providing the handle by which it may be accelerated to such speeds using magnetic fields. At this speed, relativistic effects will make the mass of the nucleus increase, while it shrinks in the direction of motion to become a flattened pancake. The two effects involve the same relativistic factor, so the mass increases to 108 times its rest mass while its thickness along the line of flight shrinks to 1/108; times the thickness of a stationary gold nucleus. In round terms, it is a hundred times heavier (with a mass of more than 100 GeV per nucleon), but the thickness of the pancake is now only 1 per cent of its diameter measured across the line of flight.

If such a relativistic nucleus meets an identical nucleus travelling the opposite way, the results will be spectacular. With (just over) 100 times as much mass in 1 per cent of the original volume for each nucleus, the density of matter in the colliding nuclei at the moment of overlap is more than 20000 times the density of an ordinary gold nucleus; the same kind of densities would be achieved in collisions involving nuclei of other heavy elements, such as lead or uranium. And as the two nuclear pancakes try to pass through one another, there will be repeated collisions between protons and neutrons meeting head on, and between nucleons and the wreckage produced by collisions that have taken place just in front of them. The best picture of what happens then comes from physicists' standard model of nucleons as composed of quarks.

Each proton and neutron — each nucleon — contains three quarks. But, as I have mentioned, quarks cannot exist in isolation. They come either in triplets or in pairs, and the best way to understand this is to think of them as being held together by a piece of elastic (in reality, an exchange of gluons) holding two quarks together. If you tried to separate two quarks, this elastic would stretch, and energy put into separating the two quarks would be stored in a way analogous to the way energy is stored in a stretched piece of elastic, or a stretched spring.

Up to a point, this means that two quarks joined in this way are held together *more* tightly the further they are apart — the opposite of the way in which familiar forces like magnetism or gravity operate. Eventually the stretched 'elastic' will snap, but only when enough energy has been put into the system to create two 'new' quarks ($E = mc^2$ again!), one on each side of the break.

The process is reminiscent of trying to separate a north magnetic pole from a south magnetic pole by sawing a bar magnet in half. Every time you break the two poles apart, you find you are left with two new bar magnets, each with a north pole and a south pole, instead of two separated poles.

So the picture of the collision between two heavy ions moving at relativistic speeds is one in which quarks are ripped out of individual nucleons, stretching the elastic joining them to other quarks until it breaks, creating new combinations of pairs and triplets of quarks, with an overlapping tangle of breaking and re-joining elastic, like high energy spaghetti. Tangled elastic may end up joining two quarks moving in opposite directions at close to the speed of light, absorbing large amounts of the kinetic energy of the collision and snapping to produce a host of new particles at the site of the collision, after what is left of the original nuclei has moved away. This is the quark-gluon plasma that physicists are eager to study — the 'little bangs' in which conditions that may not have existed for 15 billion years, since the Big Bang itself, can be reproduced. And because particles are being manufactured out of energy (the relativistic kinetic energy of the colliding nuclei), it is easy to produce a greater mass of particles from this mini-fireball than the mass of the two original nuclei. Such particle collisions are *not* simply a question of breaking the incoming nuclei apart to release their constituent components, but are a means to create the high energy densities out of which new particles can be formed. The energy to make the new particles has come from the magnetic fields used to accelerate the original nuclei.

How close are the experimenters to achieving the quark-gluon plasma? And how will they analyse the little bangs if they do succeed in creating them?

Existing particle accelerators were simply not built to do this kind of experiment, and quite apart from the energy input required there are other constraints which limit the kind of nuclei that can be used — at CERN, for example, the SPS accelerator can only be operated with nuclei that have equal numbers of protons and neutrons, while very heavy nuclei always have more neutrons than protons. Working with nuclei of sulphur-32, the SPS can reach 19 GeV per nucleon, one fifth of the way to the kind of collisions discussed here. Existing accelerators at Brookhaven can reach 5 GeV per nucleon, using silicon-28. With new booster systems (sometimes known as 'pre-accelerators'), both labs will soon be able to handle heavier nuclei, including lead, but only at about the same relativistic factors.

But by the end of the 1990s Brookhaven's Relativistic Heavy Ion Collider (RHIC) and CERN's Large Hadron Collider (LHC) should both become operational. RHIC will run at about 200 GeV per nucleon, while the LHC should reach 300 GeV per nucleon. These are equivalent to energy densities of 3 GeV per fm^3 at a temperature of 200 MeV for RHIC, and 5 GeV per fm^3 at 220 MeV for the LHC, both well in the range where, theory says, the quark-gluon plasma should form.

The way to test whether the quark-gluon plasma actually has formed is to look for a 'signature' that could not be produced in any other way. The particles that are actually detected emerging from the little bang will, of course, be everyday particles such as hadrons and electrons; but these will carry the imprint of the conditions under which they were born. One specific prediction of the standard theory is that electron-positron and muon-antimuon pairs ('thermal dileptons') should emerge from the little bangs — but the experimenters are equally interested in all products of these reactions, anticipating some surprises as they probe into the unknown. Any insight into the way everyday matter is produced out of such a quark-gluon plasma will shed new light on the way in which the kind of matter we are made of was created in the Big Bang itself.

One particularly intriguing possibility is that the mini-fireballs of the little bangs may produce a different kind of stable matter from that we are made of. Although protons and neutrons each contain three quarks, only two *types* of quark are present in everyday matter — the up (u) and down (d) quarks (and their antiparticle counterparts). A proton consists of two up quarks (each with a charge of +⅔) and one down quark (with a charge –⅓) bound together by gluon elastic, while a neutron contains two d quarks and one u.

There are, however, other types of quark, which take part in interactions at high energies and combine to form more exotic, usually short-lived, particles. One of these additional quarks, the so-called 'strange' (or s) quark should, according to theory, be able to combine with u and d quarks to form droplets of 'strange matter' (or 'quark nuggets') containing roughly equal numbers of u, d and s quarks. There have been claims that such strange matter has been detected in cosmic rays, but as yet no unambiguous evidence for the existence of such droplets has been found. But a quark-gluon plasma should produce strange droplets, as well as ordinary hadrons. They would have roughly zero electric charge (the s quark has a charge of –⅓), but masses comparable to those of a large atomic nucleus, and would be very easy to detect, providing a striking signature for the quark-gluon fireball.

Some researchers have suggested that it may even be possible to measure the size and shape of the mini-fireball itself, using a technique originally developed to measure the sizes of distant stars. The technique is named after the two astronomers who developed it, R. Hanbury-Brown and R. Q. Twiss. As used by astronomers, it involves making simultaneous observations of the same star using two telescopes up to 200 metres apart, and combining the light signals from the two telescopes to produce an interference pattern. The interference pattern then provides information about the size of the star that the light is coming from.

The Hanbury-Brown–Twiss method might be adapted to measuring the properties of the mini-fireballs produced in heavy ion collisions, using pions instead of light as the probe. The quark-gluon plasma should produce copious quantities of pions, and by studying the way in which these particles interfere with each other (remember that in the quantum world every particle is also a wave) it should be possible to infer the geometrical properties of the fireball they are emerging from.

It's worth spelling out just how breathtaking a leap in scale this involves. The Hanbury-Brown–Twiss technique was developed to measure the sizes of stars (typically 10^9 metres across) across distances of several light years. It is now being adapted to measure the sizes of mini-fireballs typically less than 10 fm (less than 10^{-14} m) across, at a distance of a metre or so. This involves a scaling factor of twenty-three orders of magnitude!

Such a trick will not be easy, but if it succeeds it will neatly close the circle of this investigation. Interest in creating the quark-gluon plasma comes from its importance in cosmology (the Big

Bang) and possibly in astronomy (the insides of neutron stars); it will be appropriate if a technique developed by astronomers can be adapted to provide insights into the nature of the bubbles of quark-gluon plasma that the experimenters confidently expect to be manufacturing before the next decade is out.

Bibliography

The books mentioned here provide useful background information on the topics I discuss in the present book. The best way to keep up to date with new developments is through the pages of magazines such as *New Scientist*, *Science News* and *Scientific American*.

Close, F., Marten, M. and Sutton, C., *The Particle Explosion*, OUP, Oxford, 1987.
   Excellent 'coffee table' book about the discovery of elementary particles.

Davies, P., *The Forces of Nature*, Cambridge University Press, 1979.
   A very clear account of the concepts underlying the modern understanding of the world within the atom — particles, fields and quantum theory. Light on maths, strong on the 'feel' of the subject. My favourite Paul Davies book and a good place to find out more about symmetry and unification of the forces.

Davies, P., *Superforce*, Heinemann, London, 1984.

Davies's best 'popular' book covering much the same ground as *The Forces of Nature* but for an audience with a less specialized knowledge of physics. Free from some of the constraints of writing textbook physics, Davies leaps off into more speculative realms in the later parts of the book and discusses subjects such as antigravity, the 'holistic' view of nature and even (in passing) astrology. If you read this and *The Forces of Nature* you'll get a good idea of what physicists are sure of, and also a glimpse of some of their wilder flights of fancy.

Feynman, R., Morinigo, F. and Wagner, W., *Lectures on Gravity*, Addison-Wesley, Reading, Mass., 1995.

Startlingly modern ideas about the way gravity fits into the quantum theory of fields, developed more than thirty years ago by one of the greatest physicists of all time.

Fritzsch, H., *Quarks*, Pelican, London, 1984.

The author is a German physicist who worked with Murray Gell-Mann on the theory that became known as quantum chromodynamics, the 'colour' theory of quarks. The book, first published in German in 1981, provides a very clear introduction to the world of particle physics, with the new theories related to the experiments being carried out at high energy accelerators in the 1960s and 1970s. It only nods briefly in the direction of the electroweak unification, and has nothing on GUTs, supersymmetry or cosmology. But if you specifically want to know about quarks, that is an advantage, since you get an uncluttered view of the main subject matter. An excellent book.

Glashow, S., *The Charm of Physics*, American Institute of Physics, New York, 1991.

'Horse's mouth' account from a Nobel Prize-winning physicist of quarks, the search for a grand unified theory, and much more besides.

Gribbin, J., *In Search of Schrödinger's Cat*, Bantam/Black Swan, New York/London, 1984.

The story of the development of quantum physics — the astonishingly successful, but non-commonsensical, theory of the world of the very small — during the twentieth century. Particle-wave duality, quantum uncertainty and the rest explained in (I hope!) readable fashion.

Gribbin, J., *In Search of the Big Bang* (revised edition), Penguin, London, 1998.

An up-to-date account of modern ideas about the origin of the Universe and the interface between particle physics and cosmology.

Gribbin, J., *Q is for Quantum*, Weidenfeld, London, 1998.
A–Z guide to the world of particles and fields.

Gribbin, J. and Gribbin, M., *Richard Feynman: A Life in Science*, Penguin, London and Plume, New York, 1997.

A life which touched on just about every aspect of physics from 1940 to 1990, including the ground covered in this book.

Ne'eman, Y. and Kirsh, Y., *The Particle Hunters*, Cambridge University Press, 1986.

The story of particle physics from the discovery of the electron and the proof that atoms are divisible (almost exactly a hundred years ago) to the evidence for the W and Z particles, in the 1980s, that suggests today's physicists are on the trail of a unified theory linking all the known particles. Just the place to find out more about quarks, baryons, leptons and neutrinos.

Pagels, H., *The Cosmic Code*, Michael Joseph, London, 1982.

A book for the general reader about the mysteries of the particle world, from a leading physics teacher.

Pagels, H., *Perfect Symmetry*, Michael Joseph, London, 1985.

Pagels' second superb book describes the search for a unified theory (as of the mid-1980s) and relates it to the best theories of the origin of the Universe.

Pickering, A., *Constructing Quarks*, Edinburgh University Press, 1984.

Subtitled 'A sociological history of particle physics', this is a book aimed at specialists, both physicists and historians of science, but is quite accessible to anyone seriously interested in following up the saga of how high energy physics studies led to the idea of quarks as the building blocks of matter. Very thorough, with copious references and scarcely any mathematics, but pulling no punches when it comes to dealing with the basic concepts, such as gauge theory and symmetry.

Polkinghorne, J. C., *The Quantum World*, Longman, London, 1984.

A very neat little book — just 100 pages, including a mathematical appendix and a glossary — that gets across most of the strangeness about quantum physics, and the philosophical discussions it has engendered.

Rae, A., *Quantum Physics: Illusion or Reality?*, Cambridge University Press, 1986.

A readable little book on the puzzles of the physics of the world of the very small — atoms and particles. As its title suggests, it deals mainly with quantum physics, but includes an excellent chapter on thermodynamics and the arrow of time.

Sutton, C., *The Particle Connection*, Hutchinson, London, 1984.

An excellent account of the work leading up to the discovery of the W and Z particles, that confirmed the predictions of the electroweak theory, written from an experimenter's point of view. This is the place to find out how the particle accelerators work, as well as getting a flavour of the excitement of particle physics in the late 1970s and early 1980s. Infuriatingly, the book has no index. Otherwise, its faults are few.

Index